Model-Based Inquiry in Biology

Three-Dimensional Instructional Units for Grades 9–12

Model-Based Inquiry in Biology

Three-Dimensional Instructional Units for Grades 9–12

Ron Gray and Todd Campbell

nsta Press
National Science Teaching Association

Arlington, Virginia

Cathy Iammartino, Director of Publications and Digital Initiatives

Art and Design
Will Thomas Jr., Director

Printing and Production
Catherine Lorrain, Director

National Science Teaching Association
Erika C. Shugart, PhD, Executive Director

1840 Wilson Blvd., Arlington, VA 22201
www.nsta.org/store
For customer service inquiries, please call 800-277-5300.

NSTA is committed to publishing material that promotes the best in inquiry-based science education. However, conditions of actual use may vary, and the safety procedures and practices described in this book are intended to serve only as a guide. Additional precautionary measures may be required. NSTA and the authors do not warrant or represent that the procedures and practices in this book meet any safety code or standard of federal, state, or local regulations. NSTA and the authors disclaim any liability for personal injury or damage to property arising out of or relating to the use of this book, including any of the recommendations, instructions, or materials contained therein.

Library of Congress Cataloging-in-Publication Data
Names: Gray, Ron, 1976- author. | Campbell, Todd, 1969- author.
Title: Model-based inquiry in biology : three-dimensional instructional units for grades 9-12 / by Ron Gray and Todd Campbell.
Description: Arlington, VA : National Science Teaching Association, [2022] | Includes bibliographical references and index.
Identifiers: LCCN 2022011560 (print) | LCCN 2022011561 (ebook) | ISBN 9781681406732 (paperback) | ISBN 9781681406749 (pdf)
Subjects: LCSH: Biology--Study and teaching (High school) | Biological models.
Classification: LCC QH315 .G73 2022 (print) | LCC QH315 (ebook) | DDC 570.71--dc23/eng/20220405
LC record available at *https://lccn.loc.gov/2022011560*
LC ebook record available at *https://lccn.loc.gov/2022011561*

Contents

Section 1 Using Model-Based Inquiry

Section 2 The MBI Units

Unit 1. From Molecules to Organisms
Structures and Processes (LS1): General Sherman

Contents

Unit 2. Ecosystems
Interactions, Energy, and Dynamics (LS2): Whale Plumes Change the Atmosphere

Contents

Unit 3. Heredity
Inheritance and Variation of Traits (LS3): Skin Cancer

Contents

Unit 4. Biological Evolution
Unity and Diversity (LS4): *Lampsilis* Mussel

SECTION 3 Appendixes

Preface

A Framework for K–12 Science Education (NRC 2012) highlights the importance of deepening all students' understanding of disciplinary core ideas and the application of crosscutting concepts through active participation in science and engineering practices. The *Framework* emphasizes that science is not just a body of knowledge but also a set of practices for developing a better understanding of the natural world. In addition, it calls for a focus on a limited number of disciplinary core ideas and crosscutting concepts explored through science and engineering practices over extended units of instruction, so students have opportunities to build knowledge and revise their understanding over time.

Currently, not enough students have opportunities to engage deeply in science and engineering practices, particularly those essential to developing science knowledge, such as developing and using models, constructing explanations, and engaging in argument from evidence. Instead, traditional science curricula emphasize memorization of discrete facts and focus on a broad range of topics. Current curriculum materials in science are not well aligned with the *Framework*. Consequently, without access to high-quality curriculum materials, teachers are left to scavenge for resources.

The model-based inquiry (MBI) framework is designed to address these issues and to provide teachers across the country with high-quality curricula aligned with the *Framework* and important ideas from research, such as the use of anchoring phenomena and public records. Additionally, visions for particular ways of thinking about engaging students in MBI are provided to support your professional learning as you prepare for and implement MBI units. In the units that follow, you will find that curriculum builds strategically from unit to unit and integrates disciplinary core ideas, crosscutting concepts, and science and engineering practices, with anchoring phenomena to help provide a context for engaging students in the application of life science ideas through the use of knowledge production practices.

Reference

National Research Council (NRC). 2012. *A framework for K–12 science education: Practices, crosscutting concepts, and core ideas.* Washington, DC: National Academies Press.

About the Authors

Ron Gray is a former middle school science teacher and an associate professor of science education in the Department of STEM Education at Northern Arizona University. He taught in South Central Los Angeles and Salem, Oregon, before earning his PhD in science education at Oregon State University in 2009. His work focuses on providing secondary science teachers with the tools to design and implement effective and authentic learning experiences for their students. Much of this work has been centered on model-based inquiry and the integration of scientific practices in a supportive and structured way. He is also interested in the history of science and science studies, which, taken together, help provide a background for understanding what "authentic" scientific practice in the K–12 context might look like. He is currently an associate editor for the *Journal of Science Teacher Education* and has published several articles for science teachers in National Science Teaching Association (NSTA) journals.

Todd Campbell is a former high school and middle school science teacher and professor of science education in the Department of Curriculum and Instruction in the Neag School of Education at the University of Connecticut. His research focuses on cultivating imaginative and equitable representations of STEM activity. This is accomplished in formal science learning environments through partnering with preservice and inservice science teachers and leaders to collaboratively focus on supporting student use of modeling as an anchoring epistemic practice to reason about events that happen in the natural world. This work extends into informal learning environments through a focus on iterative design of such spaces and equity-focused STEM identity research. He is currently the co-editor in chief for the *Journal of Science Teacher Education*. He has also published numerous articles in the NSTA journals and has served as a guest editor for a special issue of *The Science Teacher* focused on developing and using models in September 2017.

About the Contributors

Our team also consists of the following contributors, without whom the book could not have been written. All are current secondary science teachers who were intimately involved with the planning, piloting, and refining of the units included in this book.

Laurie Abo received a BA in biological sciences education from the University of Delaware, an MA in biology from Central Connecticut State University, and an MA in curriculum and instruction from the University of St. Joseph. A secondary science teacher for 19 years at East Hartford High School in Connecticut, she has participated in the UConn Mentor Teacher Collaborative for several years developing tools to help teacher candidates implement the *Next Generation Science Standards* (*NGSS*) in their classrooms. Laurie continues to seek out and participate in opportunities to learn and spread her love of student-driven, *NGSS*-based classrooms.

Audrey Baird received a BS in secondary education with a concentration in Earth science and biology and an MA in secondary science teaching from Northern Arizona University (NAU). She has been teaching secondary science, from middle school up through AP-level courses, for six years in Arizona. Audrey has been working in conjunction with NAU's Center for Science Teaching and Learning, providing mentorship to practicum students pursuing secondary science education degrees. As the department chair, she has redesigned the science curriculum for students throughout the campus to create a continuum of science concepts and skills throughout their time at the secondary level.

Jonathan Griffith works for the University of Colorado–Boulder as an education and outreach associate for the Cooperative Institute for Research in Environmental Sciences (CIRES). He works on several projects at CIRES as a curriculum developer, including the Multidisciplinary Drifting Observatory for the Study of Arctic Climate (MOSAiC) expedition. Previously, Jon was a middle school life science teacher at Mount Elden Middle School in Flagstaff, Arizona. Before becoming a classroom teacher, he was a graduate student at Northern Arizona University, where he received both an MS in geology studying lake sediments as a proxy for past climate and environmental change and an MA in science education. He is passionate about connecting students to the natural world through authentic and engaging science curriculum.

Stacy Leone received a BS in marine and freshwater biology from the University of New Hampshire. Following graduation, she went on to earn her teaching certificate at Central Connecticut State University. She has been teaching secondary science at East Hartford High School in Connecticut for more than 20 years. During that time, she received an MA in biology from Central Connecticut State University and an MA in education with a concentration in curriculum and instruction from the University of St. Joseph. Most recently, she has been working with the Mentor Teacher Collaborative at UConn, whose work is focused on supporting the implementation of the *NGSS* with preservice and inservice teachers. Stacy is focused on finding ways to engage and inspire the next generation of scientists and citizens.

Acknowledgments

We would like to thank the many science teachers, both preservice and inservice, who have worked with us and helped us refine our ideas that have culminated in this book. This work happened over the past decade, across many contexts, at Northern Arizona University, the University of Connecticut, and the University of Texas–Austin. Beyond pushing our ideas about great science teaching and learning, they helped us think carefully about model-based inquiry and the types of tools and resources that are helpful in supporting both teachers and students as they engage in more authentic representations of science in classrooms. Without them, this book would not have been possible.

Specifically, we thank the faculty of the biology department at East Hartford High School in Connecticut for their help in piloting and contributing lesson ideas as Laurie Abo and Stacy Leone collaborated with them to pilot units before the book's publication. The dedicated teachers in the biology department include Tyler Hoxley, Melissa Dumas, Jenna Rodrigue, and Lyn Douville.

Introduction

What Is MBI and Why Is It Important?

Model-based inquiry (MBI) is a framework for designing units and engaging students in science learning experiences that is focused on the construction, critique, revision, and testing of models by groups of students in science classrooms as they seek to explain events that happen in the world. This focus on explaining events in the world is important for several reasons. First, it creates a space for students to participate more authentically in the knowledge production practices of the science disciplines (NRC 2012). Second, it creates a problem space in which students can learn to use their own ideas along with the disciplinary core ideas, crosscutting concepts, and practices of science to make sense of the world around them. Third, focusing on explaining events shifts the emphasis of classroom activity from "We need to learn about this topic in order to do well in class" to "How will this help us figure out why or how something happens?" Finally, and perhaps most importantly, explaining events creates a need to learn and helps students understand why science is useful. This framework, with its emphasis on explaining events through the development of models, provides a context for the work of students in the science classroom and a way for teachers to create learning experiences that are meaningful to students.

Most students in U.S. science classrooms have not had an opportunity to explain a real-world event. Instead, students are expected to learn about many different topics in science in an abstract manner. In many classrooms, a topic is chosen (e.g., natural selection), and then students are given several different learning experiences, such as lectures, labs, and other activities, to help them understand the topic. These experiences are intended to ensure that students develop adequate knowledge about the topic before they take a summative assessment.

In contrast to this more traditional approach, MBI learning experiences, organized in instructional units, are driven by a need to explain the events that happen in the world. These events are often described as anchoring phenomena because they serve as an anchor or foundation for a sequence of classroom activities and create a reason to learn specific core ideas, crosscutting concepts, and practices of science. Student ideas about the science around an anchoring phenomenon are the central focus of each instructional unit. These ideas are elicited early, and students keep track of them with public records and collaboratively negotiate with peers throughout the unit. An MBI instructional unit culminates with students writing individual evidence-based explanations of the anchoring phenomenon. (See Figure I.1 on page xviii for a comparison between traditional science units and MBI science units.) In sum, MBI experiences are about equitably engaging all learners in meaningful forms of participation in science classrooms in ways that value the resources (e.g., ideas, experiences, knowledge production practices) that learners bring to the learning environment, while also supporting the collaborative construction and critique of explanations with these resources.

Figure I.1. Comparisons Between Traditional and MBI Units

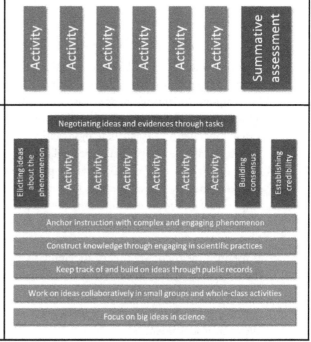

Traditional science units
Traditional science units are based on topics and activities. Once a topic is chosen (e.g., natural selection), specific activities build knowledge around the topic and are summatively assessed at the end of the unit.

MBI units
MBI units, on the other hand, are based on a need to explain the anchoring phenomenon. Student ideas about the science important to the phenomenon are at the center of the unit. Their ideas are elicited early, and students keep track of them with public records and hold collaborative peer reviews throughout the unit. The unit culminates with individual written evidence-based explanations of the anchoring phenomenon.

Where MBI Came From and Where It Is Going

In our previous collaborations over the years with preservice and inservice science teachers, we have come to appreciate the importance of modeling as a practice, both for students to work with their ideas and find connections between ideas and for teachers as the models provide insight into the ways students' explanations are evolving. Additionally, we have come to see the practice of modeling as a knowledge-building practice that is inextricably linked to other practices, such as asking questions, arguing, investigating, and explaining. Put more succinctly, it really is not possible for students to engage in modeling without engaging in the other science practices. Having noted this in our work with teachers and students in classrooms, we have collaborated with teachers and other researchers over the past decade to identify, test, and refine productive ways of engaging students in the practice of modeling across instructional units. In the end, this book is the product of that work.

How We Designed These Units

The design of the units presented here was a highly collaborative effort between teacher educators and secondary science teachers who developed, tested, and refined the units over the course of a year. The group codesigned each unit and tested it out in classrooms at least three times before creating a final version. Practical tips from our collaborating teachers are included in the units to help make implementation in your classroom as smooth as possible.

How to Use This Book

The intended audience of the book is primarily practicing high school biology teachers. While the MBI units are written toward the *Next Generation Science Standards* (*NGSS*), we believe teachers in non-*NGSS* states will find them useful as well. The four units included here are not meant to provide an entire yearlong curriculum. We chose to focus each unit on one of the four life science disciplinary core ideas (DCIs) from the *NGSS*. No unit is sufficient to cover all the performance expectations for each DCI. You can see what is and what is not included in Appendix A. In a traditional yearlong high school biology course, the units described here could be expected to cover approximately half of the required curriculum. However, the four units are designed to cover the most important ideas in each DCI.

We want to make clear that teachers may choose to substitute certain labs or activities for tasks provided here. As long as the new tasks are targeted at the same important science ideas, then students will be able to use them just as well to apply their new understandings to their building explanations of the phenomena. The effectiveness of the MBI units is not necessarily in the specific tasks. Rather, it is the role of the anchoring phenomenon, the organization of the science ideas throughout the unit, and the intellectual work done during the four MBI stages (as described in Chapter 1) as students use their ideas and those presented through tasks to collaboratively construct their final evidence-based explanations of the phenomena at the heart of these units.

Organization of This Book

This book is primarily divided into two sections. In Section 1, Chapter 1 introduces MBI, and Chapter 2 provides in-depth discussions of the ideas framing MBI and the specific stages of MBI. Section 2 contains the four complete MBI biology units. Each unit has multiple components:

- The unit summary, which includes an overall description of the unit, a summary of the phenomenon anchoring the unit, a driving question, an example target explanation, and guides for how the unit tasks fit together to lead students to their final evidence-based explanations of the phenomenon.

- Stage summaries, which give specific descriptions of each of the four stages of the MBI process. These are placed at appropriate spots throughout the unit and are a reminder that the MBI unit is more than the tasks presented.

- The tasks, each of which consists of two components:

  - Teacher Notes, which present information about the purpose of the task and explain what you need to do to guide students through it.

  - A Student Handout, which provides students with the necessary information and space for responses to complete the task. The handouts can be photocopied and given to students at the beginning of the task.

The book concludes with three appendixes in Section 3:

- Appendix A contains standards alignment matrixes that can be used to assist with curriculum planning.

Introduction

- Appendix B includes a peer-review guide that can be photocopied and given to students.
- Appendix C is a safety acknowledgment form that can also be photocopied and given to students.

Supplementary Materials

This book includes supplementary materials in the form of PowerPoint presentations, which are referenced in the appropriate places in the text. Several are step-by-step guides that lead you through the Eliciting Ideas About the Phenomenon stage for each unit; the others serve as templates that you can use with your class for the other stages and for model testing and revision. These materials can be accessed on the book's Extras page at *www.nsta.org/mbi-biology*.

Safety

Doing science through hands-on, process- and inquiry-based activities and experiments helps foster the learning and understanding of science. However, to make for a safer experience, teachers and students must follow certain safety procedures based on legal safety standards and better professional safety practices. Tasks include relevant safety precautions that will help make a safer hands-on learning experience for you and your students. In some cases, eye protection and additional personal protective equipment (nonlatex aprons and vinyl or nitrile gloves) are required, based on potential safety hazards and resulting risks. Safety glasses or safety goggles must meet the ANSI Z87.1 D3 safety standard. For additional safety information, check out NSTA's "Safety in the Science Classroom" at *https://static.nsta.org/pdfs/SafetyInTheScienceClassroom.pdf*.

Reference

National Research Council (NRC). 2012. *A framework for K–12 science education: Practices, crosscutting concepts, and core ideas*. Washington, DC: National Academies Press.

CHAPTER 1
Model-Based Inquiry

Stages of Model-Based Inquiry

Model-based inquiry (MBI) is based on a need to explain anchoring phenomena and to put student ideas and resources at the center of instruction. We believe that curriculum is necessary but not sufficient for great instruction. Curriculum provides the foundation for the important work that happens in your classroom. Layered over curriculum, however, is great instruction. As you'll see, we have purposely designed the process to provide the space and resources for effective instruction within the units. These connections are made more explicit in Chapter 2. In short, we believe in instruction that promotes equitable science instruction and the construction of students' identities in science. We see MBI as a way to transform science curricula and teaching practices to realize this vision.

MBI is broken up into four distinct stages (see Figure 1): (1) eliciting ideas about the phenomenon, (2) negotiating ideas and evidence through tasks, (3) building consensus, and (4) establishing credibility. These four stages are included in each MBI unit. Each stage consists of several activities and, as a result, will take two or more instructional days to complete. An MBI unit might be different from a traditional science unit and other instructional approaches that focus on the practice of modeling. Each stage is described in detail in the following sections.

Figure 1. The Four Stages of an MBI Unit

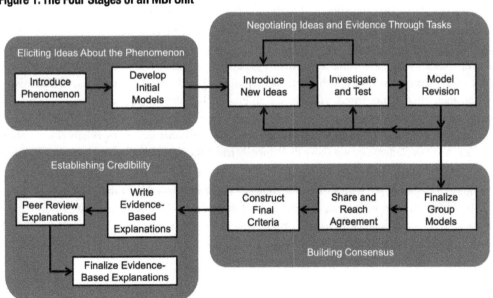

Stage 1: Eliciting Ideas About the Phenomenon

The first stage of MBI, eliciting ideas about the phenomenon (Figure 2), involves introducing the anchoring phenomenon and driving question; eliciting students' ideas and experiences, which may help them begin to formulate explanations for the phenomenon; and developing initial models based on those ideas. This phase usually takes the entire first day and includes putting the students into groups of three or four that will work together throughout the unit.

Figure 2. Eliciting Ideas About the Phenomenon

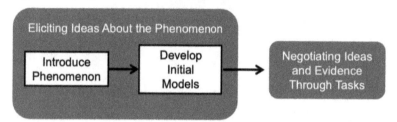

We begin this stage by introducing the phenomenon in an engaging way, such as with stories, videos, demonstrations, or even short activities. The goal is to provide just enough information for students to begin to reason about the phenomenon, without providing too much of the explanation. While we are introducing the phenomenon, we ask questions to keep students engaged and make sure they are paying attention to the important aspects. The introduction ends with the driving question of the unit, which we have found helps focus their thinking on the development of a causal explanation for the phenomenon, or a "why" answer. We then get students into their groups and facilitate the first discussion to try to answer the driving question with just the resources they brought with them—the ideas, experiences, and cultural resources they have gained both in and outside of the classroom. These ideas may be fully formed, partially correct, or fully incorrect in terms of our canonical knowledge of science. However, we make it clear that all ideas are considered equally valid at this point in time, as we realize that these are the ideas that students put into play when they think about the phenomenon we have introduced. As described more fully in Chapter 2, it is these ideas, along with those we introduce through the tasks, that are at the heart of the sensemaking that students need to do to explain the phenomenon.

Once student groups have discussed their ideas, we facilitate a class discussion to compare and contrast the ideas generated by each group. As ideas are presented, they are put on our first public record, which we call the Initial Hypotheses List. We use this list throughout the unit to keep track of the changes in students' thinking as they work toward a final evidence-based explanation for the anchoring phenomenon. We consider all ideas to be valid at this stage, before students use other resources to begin making sense of how or why something happens. The Initial Hypotheses List is also useful for the next task in this stage, initial model construction, especially since it offers students additional ideas beyond those they initially had either individually or in their small groups.

If students are not experienced with modeling, it is worth providing a brief introduction and example. Examples of modeling are included in the Eliciting Ideas About the Phenomenon PowerPoints for each unit, which can be downloaded from the book's Extras page at *www.nsta.org/ mbi-biology*. Once the class is ready to begin modeling, we give each group a sheet of 11 × 17 inch paper and ask the groups to each make a model of their initial hypothesis. Sometimes they choose their own original hypotheses, while other times they are influenced by their peers' ideas and adopt one of them instead. As the groups work on constructing their models, you should walk around asking clarifying questions and pushing students to be as specific as possible. Once the models are ready, it is important to have students share ideas across them. There are a number of ways you might run these share-out sessions. We often collect and present the models on a document camera at the end of the first day. Groups can provide one- or two-sentence summaries of the initial hypotheses that they have represented in their models. We point out interesting ideas and ways in which they have represented these ideas. For example, we may call attention to the fact that a group labeled the arrows, which made the model more understandable, and that another group used a zoom-in window to show what was happening at a different scale. At the end of the first day and this first stage, we have elicited ideas across the class, and the groups' initial hypotheses and models will act as a starting point for the rest of the unit.

Stage 2: Negotiating Ideas and Evidence Through Tasks

The goal of the second stage, negotiating ideas and evidence through tasks (Figure 3), is to support students' ongoing changes in thinking by providing learning experiences that help coordinate their ideas, core ideas, and crosscutting concepts to build a scientific explanation of the anchoring phenomenon. This involves designing or adapting a number of data-based tasks, introducing core ideas and crosscutting concepts, and constructing and using public records such as a Summary Table to help keep track of what students have figured out, how they know what they know, what these ideas help them explain, and new questions they may have. Important in this stage are the revision and testing of students' models. This stage makes up the majority of the unit, as the class works to develop explanations of the phenomenon through engagement in the practices of science. For each important idea, there is at least one task, and the steps shown in Figure 3 and described in this section are repeated for each task.

Figure 3. Negotiating Ideas and Evidence Through Tasks

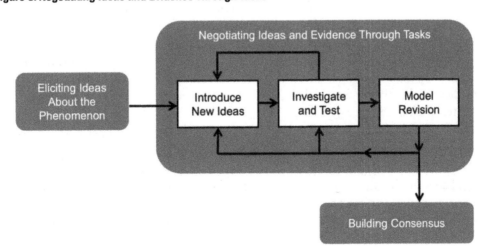

Throughout this book, we prioritize a basic sequence of practices before, during, and after each task. These are flexible depending on your context, but we have found them to work well based on our goals of equity and using students' ideas as resources for sensemaking. First, instead of relying on students to discover important science ideas from our tasks, we highlight these ideas before the task begins. This often happens with short, direct instruction in which we highlight an idea such as natural selection. Then the tasks engage students with data that link to these ideas as part of a sensemaking experience. We don't see this as giving away the answer. Instead, it allows them more time to reason about the idea before ending the task by doing the intellectually challenging work of applying it to the phenomenon. The true intellectual work of a task is not just understanding the concept of natural selection, for example, but figuring out how to apply that concept to explain some event in the world, which in the task is the anchoring phenomenon.

While student groups are working on the task, we engage with each group and ask back pocket questions that are designed to stretch their thinking. This usually involves going to a group, listening in for a minute or two to understand how their conversation is progressing, and asking questions that press them to go further. We also focus on making sure each person in the group is able to share their ideas. As we finish, we ask a "leaving question" that encourages students to continue thinking about what they are doing after we have left for the next group. This process also allows us to monitor where groups are in their thinking so we know whom to call on for the most productive discussion after the task has finished.

After students have completed the task and put away the materials, we move to a whole-class discussion to make sense of the task as a class. This can occur in a number of different ways, depending on the nature of the task. For example, data from each group can be combined to begin to make sense of the foregrounded ideas. We use another public record, the Summary Table, after each task to wrap it up in two ways. First, it allows students to summarize what they learned from the task about the foregrounded ideas. Second, it prompts them to think about how their new ideas and understandings are related to the explanation of the phenomenon. We have found the conversations the class has while working to fill out the Summary Table are incredibly important but also very challenging. We have found ways to make filling out the table go more smoothly. For instance, we don't use bullet points but instead write one or two complete sentences on the table. To make it easier to construct these sentences, we ask that groups first write their ideas in the section titled Some Useful Ideas From My Teacher, found near the end of the Student Handout, in one or two complete sentences as well. Then, as we elicit responses and compare and contrast them, they are already fully formed thoughts, making it easier for students to summarize them in full sentences on the Summary Table.

About halfway through the unit, students should go back to the initial models constructed on the first day and revise them based on what they have learned. This can be done in multiple ways. For instance, they can review their initial models and use sticky notes to flag anything that should change based on what they've learned as documented on the Summary Table. Alternatively, they can have discussions about what to change and then redraw their models. We make the choice between these options based on the amount of time we have for the model revision process. It is important, however, that students make decisions about revisions based on what fits with the evidence from the tasks and is consistent with the scientific ideas at play.

In the end, during this stage, students have built on their initial ideas elicited on the first day by engaging in data-based tasks, each designed around one or more central scientific ideas required for a complete scientific explanation of the phenomenon. Some of these ideas were introduced by students, and we introduced others. By keeping these ideas and the ways they help build an explanation of the phenomenon at the forefront of the work in our classrooms, we help the students, working together in groups and as a whole class, make sense of these scientific ideas and how they apply to the phenomenon. As this stage wraps up, we now have all the pieces we need to come to a complete evidence-based explanation of the phenomenon.

Stage 3: Building Consensus

The third stage, building consensus (Figure 4), is about pulling it all together. Throughout the unit thus far, students have worked together in groups to make sense of the tasks in the context of the anchoring phenomenon. Through discussions and share-out sessions, we have worked to coordinate their ideas so groups could learn from each other. In this stage of an MBI unit, the whole class works to build consensus about the explanation of the phenomenon by finalizing the groups' models, comparing and contrasting those models as a whole class, and constructing a consensus checklist of the ideas and evidence that should be a part of students' final evidence-based explanations that make up the summative assessment of the unit.

Figure 4. Building Consensus

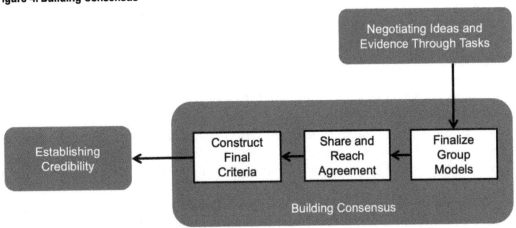

As with the model revision that occurred halfway through the unit, finalizing the models requires the groups to review their previous models, decide what needs to be revised based on new ideas and understandings from the last set of tasks, and redraw the models so they can be more easily shared and used to build consensus in a whole-class setting. This usually takes about 30 minutes. We often scaffold this process by asking students to review the completed Summary

Table and talk about what should be added to, removed from, or changed in their previous models. They should think about what new ideas can help them move toward a truly causal model that includes not only *what* happened but also *why* it happened. This requires that the mechanism at play be visible and well explained. We push the groups to make sure that items in their models are labeled, that any unseen components are made visible, and that the important ideas that surfaced throughout the unit are explicitly used in the models. There is a challenging tension here in that we want to provide guidance and encourage students to make the most complete and useful models they can, but we don't want to provide so much guidance that the process of modeling loses its power and they just create the models that we want them to. As they work on this process, we walk around and press them for more detail and clarity, focusing on the evidence that does or does not support the various components and relationships in their models.

We think that a share-out session is needed again here so that the groups can learn from each other and begin to build consensus across the models. There are a number of ways to do this, including gallery walks. As the goal is to build consensus as a whole class, we usually opt to facilitate a share-out session in which each group comes to the front of the class, displays its model, and talks it through. We prompt students to ask questions, and the class works hard to compare and contrast ideas across the models. As the unit is near the end, we push harder here than we did earlier in the unit when we wanted to allow groups to have some ideas that were still not canonically correct or completely refined. We also want to encourage students to think about how well each aspect of the model fits with the evidence they generated during the unit and whether it is consistent with the core ideas and crosscutting concepts that were introduced during the previous stage of the unit. However, our goal is for students to do this work themselves.

At this point, the role of the models is over. They are not used for a summative assessment. Instead, they were just tools to help a group of students make sense of the ideas highlighted in the tasks and how they applied to the phenomenon. Now we shift to facilitating a discussion and building a new public record around our group consensus of the evidence-based explanation of the phenomenon. Some teachers like to build a whole-class model together in front of the class here. We think this works best for younger students and tend to move right to the construction of the checklist with our older students.

The goal of the final checklist is to have the class negotiate about which of the main ideas and evidence should be part of a complete and scientifically defensible explanation of the phenomenon. Students will use this as a scaffold for writing their final evidence-based explanations in the last stage of the unit. There are a number of ways to facilitate this discussion and creation of the public record. We like to prompt groups to create a bulleted list of the three to five most important ideas they think they need. We then ask for examples, press the whole class to make sure they understand how the idea fits in, and ask for consensus before writing it on the final checklist public record. This process can take about 15 to 20 minutes and most often goes quite smoothly by this point in the unit. However, this is also a time to make sure important ideas are included and to bring them up if necessary. While this is uncommon, we may also provide some just-in-time instruction to tie up any loose ends in the students' understandings from the unit.

The checklist becomes less useful to students if everything that comes up in this discussion is automatically written on the board. Ideas can generally be combined into five to seven bulleted points. We then go back and lead a discussion about the evidence we have for each of the bulleted points and write those alongside. This step is crucial, as students will need to coordinate the science ideas with evidence in the written evidence-based explanation in the next stage. At times, we have been unhappy with our checklist for some reason; perhaps the writing is not as legible or the points are not as clearly articulated as we would like. In such cases, we created a clean version that evening and asked the students the next day to ensure that it still represented all the ideas from the original poster.

Stage 4: Establishing Credibility

Science does not progress just because a great explanation for a phenomenon has been developed. We have to argue for our explanations and convince others that they are valid before our ideas have credibility in the scientific community. Similarly, in MBI, students must argue for their ideas in writing to convince their peers and teacher that their explanations of the anchoring phenomenon are scientifically valid. In the fourth and final stage, establishing credibility (Figure 5), students do this through written evidence-based explanations, peer review, and revision. In MBI, their explanations are not just handed in for the teacher's eyes only. Students engage in conversations with peers about the strengths and weaknesses of their arguments as they work to improve their final products. In this way, the revisions provide another opportunity for students to learn from one another as they consider and critique their peers' explanations.

Figure 5. Establishing Credibility

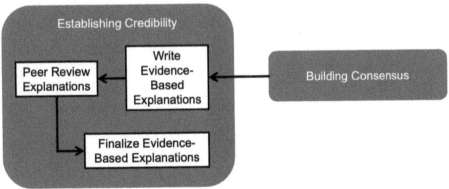

This chapter has explained the basic concepts of MBI and its four stages, eliciting ideas about the phenomenon, negotiating ideas and evidence through tasks, building consensus, and establishing credibility. We hope this summary of MBI has sparked your enthusiasm to read further about how to apply this teaching method in your own classroom, using the example units we provide.

Reference

National Research Council (NRC). 2012. *A framework for K–12 science education: Practices, crosscutting concepts, and core ideas.* Washington, DC: National Academies Press.

CHAPTER 2

Framing, Tools, and Routines for Supporting MBI Instruction

This chapter builds on the vision for MBI laid out in Chapter 1 by framing what we have found to be the most productive stances toward teaching and learning in MBI units. This is important because the way people think about what they are doing influences the kinds of resources they may consider drawing on or the tools or strategies they may decide to use as they work to explain complex events that happen in the world. In this chapter, we provide a framework related to specific features of classroom environments (e.g., student ideas as resources, modeling, explanation, arguing from evidence) from both the teacher's and students' vantage points. Often, these two vantage points converge, as they are descriptive of similar activities; however, it is important to distinguish here between how the teacher and students might need to think about each of the different features of MBI. In the MBI learning environment, your role as a "learned other" responsible for meeting curriculum expectations for learning in biology, while concurrently supporting students' engagement in authentic forms of sensemaking, can at times be different from the sensemaking role of students, who are focused on figuring out how to explain an event that happens in the world.

Given this aim, this chapter begins with some general principles for framing both teachers' and students' thinking about the MBI learning environment, including ways to think about equitable access and participation and student ideas as resources. Then these ideas, along with other stage-specific framings, are considered across the arc of stages of an MBI unit.

Equitable Access and Participation

Central to science teaching and learning is a focus on equitable access and participation, in terms of both who has access to high-quality teaching and learning materials and the particular roles, responsibilities, and expectations of learners in connection to their interactions with teachers, peers, and learning materials in science classrooms. Importantly, equitable access and participation is intricately connected to who is seen as capable of shaping the roles, responsibilities, and expectations in relation to what it means to participate. Much of what has historically happened in schools has created barriers to both equitable access and participation in science classrooms, particularly for minoritized populations. Since we see learning as participating in valued pursuits and involving more than simply acquiring knowledge, access and the ability to participate are important. While acquiring knowledge is still important for learners (and educators), it needs to be accompanied by learners' own application of knowledge in contexts they have identified as meaningful.

Beyond providing space for transformative forms of participation negotiated through the reshaping of roles, responsibilities, and expectations, allowing for equitable access and participation supports learners in drawing on and making connections between their useful experiences and ways of solving problems that are connected to their interests and community. Further, participation in meaningful pursuits can help students begin to see themselves in imaginative ways, as they are recognized by others and recognize themselves or identify as the kind of person who can do science, especially as their participation intersects with their other interests and ways of thinking about themselves (e.g., as a maker, a dancer, or an athlete).

For the **teacher**, this means you need to draw on strategies that both support equitable access and participation by providing learners with agency in shaping their roles, responsibilities, and expectations in classroom pursuits (e.g., through supporting students in proposing hypotheses that can be tested through student-designed investigations) and elicit learners' useful experiences and ways of solving problems that are connected to their interests and community (e.g., through providing early opportunities for students to express their ideas and use problem-solving strategies to resolve their uncertainty about events that happen in the world).

For the **students**, equitable access and participation means that they experience a feeling of belonging within a classroom community, with both their teacher and peers. In an MBI unit, students investigate a complex anchoring phenomenon (e.g., possible connections between whale plumes and global climate change) through sharing their ideas and questions and engaging in knowledge production practices (e.g., argument, investigation, modeling) to resolve uncertainty as they work toward community-negotiated evidence-based explanations. During this process, their feeling of belonging is related to the extent to which they are able to engage collaboratively with their peers, allowing them to feel connected to and invested in the pursuits of the community.

Student Ideas as Resources

As discussed in Chapter 1, we see student ideas as resources and prioritize the refinement of these ideas or resources in classroom communities using science practices. From a resources perspective, ideas are considered as fine-grained pieces of knowledge connected to everyday ways of thinking about the world that have been shaped by experiences and community. Viewing students' ideas from this perspective emphasizes how they can make sense of new situations using their existing ideas as stepping-stones. Student ideas include partial understandings and nonstandard ideas that are connected to different types or pieces of evidence they have collected during their everyday lived experiences. In this view, students activate their ideas and experiences they think will be helpful for developing explanations or solving problems in the particular contexts (e.g., the social and physical environments) in which they find themselves.

A Framework for K–12 Science Education (NRC 2012), subsequently referred to as the *Framework*, uses this resources perspective to prioritize sensemaking in advocating for a new vision for science teaching. Here, sensemaking can be understood as working on and with ideas—both students' ideas and authoritative ideas in texts and other materials—in ways that support students in making meaningful connections. The goal is for students to engage in knowledge-building practices (e.g., modeling, explanation, argumentation) as they use their ideas and developing understanding of core ideas or crosscutting concepts to make sense of phenomena. In this view, knowledge-building practices are tools the classroom community uses to work at knowing or to recognize when an idea is or isn't productive in the context in which it is being used.

With a resources perspective, teachers move away from a focus on misconceptions and deficit framings of students' ideas, which might prevent students from willingly sharing and working with and on their own and others' ideas in science classrooms. This is done during MBI as you orient to student ideas and sensemaking repertoires and recognize that learning necessarily entails connections between what learners have already come to know and the new ideas and available evidence they are learning about, allowing them to refine their ideas. This way of thinking about students' ideas in relation to the development of an explanation for an event is depicted in Figure 6. As the figure illustrates, early in MBI units, student ideas are elicited, and disciplinary core ideas (DCIs) and crosscutting concepts (CCCs) become more prominent in the evolving explanation of the anchoring phenomenon through the engagement in science and engineering practices (SEPs). In the later stages of MBI, student ideas are intricately connected to relevant DCIs and CCCs as a result of the consensus building and establishing credibility stages of MBI.

Figure 6. Student Ideas in MBI

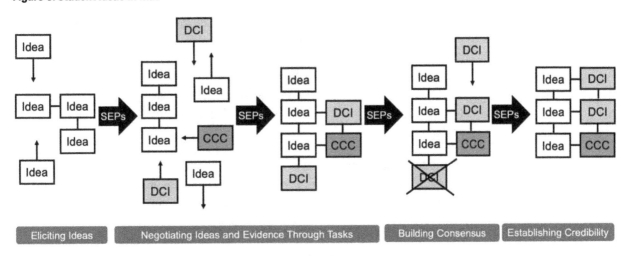

An important aspect of putting students' ideas at the center of instruction is the use of public records, which we define as representations of what we know at the time that are available for all members of the classroom community to use as they negotiate ideas over time. Public records keep ideas front and center, and they allow for the public revision of ideas. They are instrumental in the flow of ideas across a unit. In MBI, we recommend using a number of different public records throughout each unit, including Initial Hypotheses Lists and Summary Tables.

From the **teacher's** vantage point, students are trying out ideas in an MBI classroom. These ideas are often fine-grained, and it is the proposed connections between the fine-grained ideas and the evidence that supports or challenges each one that will be refined over time as students consider competing ideas and connections in building a logical and evidence-based explanatory account of a real-world phenomenon.

From the **students'** vantage point, they are trying to explain something that is happening in the world. Their job is to propose and refine ideas, recognizing that the ideas proposed in their earliest attempts represent "first drafts" of their thinking and may be refined or discarded as they also consider ideas from others (e.g., peers, the teacher, texts). Note that ideas that are refined or discarded during MBI are those that do not work as well as other ideas for explaining the anchoring phenomenon or do not fit with all the available evidence. This does not mean that these ideas are wrong or not useful. The ideas or ways of thinking that students share are valuable in many different contexts, and it is important to ensure that they continue to see value in their ideas and ways of knowing so they do not feel that they have to give up who they are or what they hope to be in order to participate in the construction of scientific knowledge.

Students at this stage, because of how their teacher presents what is going on, also begin to understand that they can propose an idea that is just something to think about or to see if it may be helpful. Further, students should understand that they will be supported to think about the idea with peers. Additionally, in MBI, the testing of ideas happens as students use the practices proposed in the *Framework* alongside locally developed practices the class has found helpful in discerning whether an idea is useful by considering the extent to which it is supported by existing or emerging observations and the available evidence.

Ways of Thinking in the MBI Stages

The following sections consider how these ways of thinking about equitable access and participation and student ideas as resources, as well as and other ways of thinking, factor into each of the MBI stages.

Eliciting Ideas About the Phenomenon

Figure 2 in Chapter 1 (p. 2) reveals how the first stage of MBI, eliciting ideas about the phenomenon, is focused more generally on introducing the anchoring phenomenon and developing initial models.

From the **teacher's** perspective, this stage is critically important, as it orients students to the MBI unit anchoring phenomenon using the driving question in a way that makes it compelling enough to spend several weeks across a unit working on explaining an event. Put more succinctly, this stage of MBI introduces the problem space in which the students will be working. It is a carefully selected phenomenon (e.g., whale plumes, *Lampsilis* mussel) that is accessible to learners and rich enough that many connections between ideas and evidence is necessary. In other words, introducing the phenomenon should be done in a way that supports students in drawing on what they have learned previously in and outside of school as they begin to think about how to explain the driving question of the MBI unit. In this stage of MBI, your priority should be to elicit students' ideas about the phenomenon, since it is important for them to continually think about the phenomenon and refine their everyday ways of thinking, especially as they are introduced to powerful disciplinary core ideas and crosscutting concepts that have been refined over time by the scientific community and proved useful to explain a range of real-world phenomena.

Since eliciting students' ideas is a top priority in this MBI stage, it's important to create a learning environment where students feel comfortable and are invited to offer ideas. This means that you need to think carefully about how to get students to float (or put on the table for consideration) as many ideas as possible. Table 1 reveals strategies we have found useful for opening up space in early small- and large-group discussions for ideas to be considered.

Table 1. Strategies for Opening Up Space for Student Ideas

Strategy	Rationale
Start with observations	Having students share what they observe ensures there are accessible entry points for all students.
Encourage use of everyday language	Encouraging students to share things in their own words, in ways they normally talk, allows them to focus on ideas instead of the accuracy of language use to connect to ways they may have already articulated ideas outside of school with friends and family.
Provide sentence stems that lower risk of sharing ideas	Sentence stems like "I think this might be caused by _____" or "I think this might have something to do with _____" can provide space in language use through words like *might*, so students feel more comfortable contributing ideas.
Connect to students' experiences	Encouraging students to think about how what they are reasoning about in the classroom might be like something they have experienced or observed outside of school provides them with space for making more connections and exploring more thinking resources (i.e., their own and their peers' ideas).

Through drawing on the strategies outlined in Table 1, you guide your class to move from observations and questions to early explanations that are further negotiated. This process of introducing the phenomenon strategically to elicit a range of ideas and the development of initial models is important, since it serves as the first step in the proposal and refinement of explanations, the ultimate aim of the MBI unit specifically and science more generally. In MBI, we think of models as the products of students' work for this stage that are intricately connected to their evolving explanations, while we think of modeling as the knowledge-production practice in which students engage to refine their thinking. We propose that you think of modeling as a practice that supports students in working at knowing, and of models as a product that, early on in the MBI unit, can support cross-group sharing and negotiation, which are accomplished through another important science practice in the *Framework*, argumentation. In this context, argumentation can be thought of as another science practice in which students engage to think about and draw on parts of their models in conversation with other groups in the classroom to persuade others or be persuaded about ideas that are not yet resolved.

Beyond thinking about modeling and models as a way of working at knowing, you can also think of models as public representations that students create that allow them to more formally propose what they are motivated to share at this early stage of the MBI unit. In this context, you can consider models and modeling as a formative way to see your students' reasoning, which might not be possible without the product of modeling (i.e., models). Figure 7 is an early model from the whale plume unit included in this book that helps reveal the power of initial models for supporting teachers in early efforts at formative assessment (i.e., collecting information about student thinking, considering what it means, and deciding how it can inform next steps of support you might offer to build on what students have already accomplished).

Figure 7. Initial Whale Plume Model

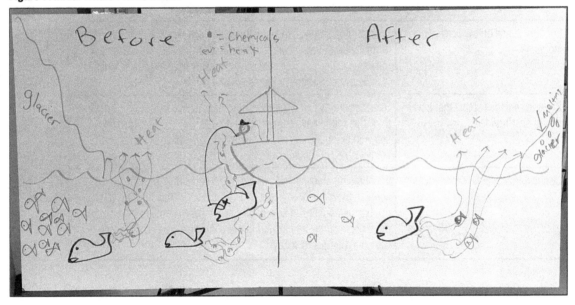

From the **students'** perspective, as they engage in this stage of MBI, they should be thinking about what they are doing as they are figuring out something that happens in the world that has been presented in such a way that it is puzzling and interesting enough to pursue, and about which they have some early ideas they can draw on that might help them contribute to the class's attempts to work collaboratively as students' ideas are proposed and refined across time. Note that this is quite different from the goal of knowing or understanding a topic. We want students to feel comfortable enough to think about the sharing of their initial ideas as important and productive for helping the classroom community make progress in its explanatory pursuits. When it comes to engaging in the practice of modeling, we want students to think of their work in collaborative groups of three or four as a sensemaking endeavor that will, because of the negotiations they engage in, lead to the groups making progress in their attempts to answer the driving question that has oriented them to the unit anchoring phenomenon. These negotiations occur both in proposing and consolidating their own ideas and sharing and considering which ideas should be included in group models and in what ways.

While it may not be necessary to have students spend time thinking about differentiating the practice of modeling and their resultant group models as outlined for teachers, we do believe it is important for students to see the act of modeling as a way they can work in small groups to propose, develop, and refine explanations of real-world events. In other words, models are tools that help a group of people (e.g., students, scientists) think together as they develop their scientific explanation for the anchoring phenomenon.

Negotiating Ideas and Evidence Through Tasks

Figure 3 in Chapter 1 (p. 3) outlines the important features of the second stage of MBI, negotiating ideas and evidence through tasks. This stage is focused more generally on refining the initial explanations students came up with in the first stage and makes up the bulk of their experiences in an MBI unit. This stage of each MBI unit includes multiple tasks, with each task specifically designed to introduce needed ideas that will be important for explaining the unit anchoring phenomenon and help students generate the evidence needed to support, challenge, and refine these explanations. During this stage, the teacher supports students in refining their initial explanations as they engage in carefully designed tasks intended to help them learn about disciplinary core ideas and crosscutting concepts that will be useful in supporting their pursuit of explaining the unit anchoring phenomenon. Any scientific explanation is built on scientific ideas.

In this stage of MBI, **teachers** should think of their role as that of helping introduce ideas to students to support them in learning about and beginning to engage with the ideas through carefully designed sensemaking tasks. If important ideas are not introduced by the students, you can put them on the table or introduce them through the use of just-in-time direct instruction, short videos, or readings, among other strategies. You can think of your role in just-in-time instruction as that of introducing new ideas. Since the objective is to have students pick up and try out these ideas in carefully crafted tasks as they develop explanations, introducing your students to an idea early in the task is *not* giving them answers that will be confirmed in the task. Instead, you can think of this as introducing students to tools (i.e., disciplinary core ideas) that will be useful in a task in ways that will help them make sense of data, a simulation, an activity, or something in the world.

In the end, since the objective of tasks is to support students in their early attempts to make connections with newly introduced ideas that are useful in completing the tasks and ultimately in making connections to the unit anchoring phenomenon, it is important that you think of introducing new ideas in the early parts of this MBI stage as an opportunity for students to engage in sensemaking with new ideas to use, while you concurrently recognize the potential explanatory power of the newly introduced ideas. When thinking about engaging students in investigating and testing the introduced ideas, you can consider these investigations or activities as the medium or context in which the introduced ideas reveal their usefulness as the planned uncertainty within the task is resolved through connections students make with introduced ideas in collaborative small-group work and large-group discussions. At the end of each task, you can look at the completion of a row of the Summary Table as an opportunity for students to think about what they have learned in the task and, together with peers, reason about how the newly introduced ideas can be applied to support their objective of explaining the unit anchoring phenomenon.

From the **students'** perspective, they should be thinking about their engagement in tasks as the opportunity to explore resources and engage in investigations and activities that will help them resolve some aspect of the important mechanistic questions that need to be answered (e.g., where do seeds get energy?) before the classroom community decides it has figured out or answered the driving question about the unit anchoring phenomenon. As students are introduced to new ideas, they should see these ideas as capable of resolving some questions that were not previously resolved with the early ideas that they and their classmates considered in their initial explanations. Students should see engaging with classmates in investigations and tasks as rich opportunities to learn with and from their peers in ways that will be beneficial.

As students increasingly think about connections between introduced ideas and investigations, they can expect to consider a broader range of connections that they can test to reveal the unique and promising connections that can be made, with the range of connections made in the task eventually supporting their attempts to apply the introduced ideas to explaining the unit anchoring phenomenon. And as students engage at the end of each task in completing a row of the Summary Table, they should see this as their opportunity to resolve some puzzling aspect of the anchoring phenomenon. Finally, in this stage of MBI, as students revisit their initial models at least once, they should see this as their chance to again engage in modeling with peers in their groups as a way of supporting them in refining previously included ideas and applying new ideas to make additional progress in refining their explanations of the anchoring phenomenon.

Building Consensus

Figure 4 in Chapter 1 (p. 5) shows the third stage of MBI, building consensus. Broadly speaking, this stage of MBI focuses on revisiting and refining group models based on new ideas and evidence, as well as sensemaking experiences, that took place in the second stage. While the second stage of MBI involves revisiting the groups' initial models at least once, this happens midway through the planned tasks and often is facilitated with sticky notes that groups use to remind themselves of changes they believe are needed based on what they learned in early tasks. Since students have completed additional tasks after revisiting the initial models, further revisions are needed at the beginning of this third stage. Beyond revising and finalizing group models, this stage also involves groups sharing and reaching consensus in relation to how their models explained the unit anchoring phenomenon, as well as the construction of final criteria or whole-class mapping of ideas included in the Gotta Have Checklist to the evidence collected from the various tasks completed in the second stage.

In this stage of MBI, it is important for **teachers** to revisit the earlier discussion of how to think about modeling and models in the first stage. In short, we described the science practice of modeling as a knowledge-building practice and models as the products of modeling. Students begin this stage by finalizing their models in groups of three or four, which you should think of as a continuation of the sensemaking that started in the groups' initial modeling experiences in the early days of the MBI unit and continued when they revisited these models midway through the second stage. As with these other experiences, you can expect the practice of modeling to support groups in refining their explanations of the unit anchoring phenomenon, so that negotiation and argumentation lead to refined ways of thinking about the phenomenon that may not have occurred had students not engaged in modeling this final time. The groups' models will offer insight into your students' thinking that otherwise may not be accessible to you. At the same time, models as products of modeling also serve as artifacts that students can compare across groups in ways that are supported by your selection and sequencing of groups' sharing in whole-class sensemaking, so that differences can be foregrounded, negotiated, and resolved through argumentation, among other science practices.

In this stage, you can also think of the construction of final criteria or whole-class mapping of ideas included in the Gotta Have Checklist to the evidence collected from the various tasks completed in the second stage as a scaffold that will be important for individual students in writing evidence-based explanations as the initial activity in the next stage of MBI. Importantly, the teacher-facilitated, co-constructed Gotta Have Checklist, which outlines what the class agrees should be included in students' explanations and is mapped to evidence identified across the unit tasks from the second stage of MBI, should be considered as the last stage of building consensus. This is especially true because final group models will have been negotiated in both small and whole groups as a basis for the identification of the criteria that are needed for writing an evidence-based explanation.

From the **students'** vantage point, they should see this stage of MBI as a time to pull the ideas and evidence they have engaged in thinking about across the unit together with those of their classmates and consider them in relation to how they reasoned with ideas in the first stage and revisited their models to add ideas connected to evidence from tasks in the second stage. As they begin this third stage of MBI, remind them that you expect that through engaging in working to finalize their models with their groups, this science practice will help them clarify their thinking and finish their engagement in modeling, having made progress as a result of their work. As they transition into sharing their ideas represented in their models with other groups, this process will also support the entire classroom community in moving forward as a collective in figuring out how to best explain the unit anchoring phenomenon.

Finally, after students have come to a consensus across groups in the whole-class setting, be sure that they understand why they are being asked to co-construct a Gotta Have Checklist that outlines what the class agrees should be included in the evidence-based explanations (because it will help them as they are asked to write their own evidence-based explanations using what they have come to understand across the unit). The goal is to ensure that students see all they are asked to do as a meaningful and important part of making progress in explaining the anchoring unit phenomenon.

Establishing Credibility

Figure 5 in Chapter 1 (p. 7) shows what this last MBI stage, establishing credibility, entails. In this stage, students initially propose a final evidence-based explanation using the criteria established at the end of the third stage. Previously, their work had been done both in small groups and as a class. Now, however, students write their evidence-based explanations individually. After they have drafted their explanations, the teacher supports them in engaging in peer review (a guide with criteria is provided in Appendix B). Students then consider the feedback they received during peer review to finalize their evidence-based explanations as the outcome of the work they engaged in across the MBI unit. This allows each student's final evidence-based explanation to gain credibility through negotiations with peers.

From the **teacher's** perspective, writing evidence-based explanations is a sensemaking experience (i.e., constructing explanations) that students engage in early in this stage of MBI. You should think of this as another knowledge-building opportunity that, like engaging in modeling earlier in the unit, is a practice that leads to a product. And like models, which served as products that could provide you with insight into the way groups were thinking about explaining the anchoring phenomenon, the individually written evidence-based explanations afford you with an opportunity to assess individual students in terms of where they are in their attempts to explain the anchoring phenomenon and how you might further support them across the remainder of this stage. As students engage with their peers to review each other's evidence-based explanations, you can think of this as additional space for argumentation and negotiation as students recognize possible differences and details they may not have considered or may not yet agree with. This practice of peer review needs to be supported, so we have provided the MBI Explanation Peer-Review Guide in Appendix B, which students can use to provide feedback to each other.

The guide contains questions designed to focus student attention on the most important aspects of explanations without being a checklist and to encourage students not only to discuss what they figured out and how they know what they know but also to best communicate what they have learned to others in a way that is clear, complete, and persuasive. During the peer review, you should consider and make explicit how explanations, like models, are refined over time through negotiation within a community. In the scientific community, groups of scientists engage in the refinement and negotiation of explanations of events that happen in the world through informal discussions, conference presentations, and journal publications. In the science classroom, students are supported in the peer review included in this stage of MBI in similar ways through engaging with their peers, who have also invested considerable effort in reasoning about the same phenomenon. The MBI unit culminates at the end of this stage as students use the feedback they have received to make final revisions that represent how they are thinking about and capable of explaining the unit anchoring phenomenon. The final evidence-based explanations serve as one measure of what students have learned about engaging in science practices to use disciplinary core ideas and crosscutting concepts to explain the phenomenon. You should also use the peer review guide in Appendix B to evaluate students' final evidence-based explanations.

From the **students'** perspective, this stage of MBI is about constructing their own individual evidence-based explanations using what they negotiated in their groups as their model for explaining the anchoring phenomenon, along with the criteria they negotiated with their teacher and peers. It provides them with a chance to see how their thinking may be similar to or different from that of their peers. You should continue to support building consensus as students negotiate any discrepancies or conflicting ways of reasoning they may identify in relation to their peers' evidence-based explanations. Help students think of this stage of MBI not as a summative test of their way of explaining the phenomenon but rather as a way to work closely with their peers to support one another in refining their explanations, which in the end will represent where they and their classmates are in respect to their current thinking about how best to answer the driving question. It is also important that students see the experience of review and revision as a useful and normal part of work in science that helps ensure consistency in how ideas and evidence are coordinated in logical ways as explanations are developed.

This chapter has built on the description of the stages of MBI outlined in Chapter 1 by considering the most productive ways teachers and students can think about the work they are doing across MBI units. The information in these two chapters should help you understand and be ready to engage learners in the units that follow. In this chapter, we have attempted to take you behind the scenes to support you in thinking about the complex work you will undertake as you implement MBI units. The way you think about your work and the way you support students in thinking about the work they are doing has a profound influence on the way they will experience learning in your science classroom. If students think they are working to understand what you already understand, and that their job is to learn from you and ask questions to elicit your knowledge, the learning experience will be profoundly different from the way students learn through MBI experiences, which position them within the classroom community to figure out how to explain something in the world. The outcome is not discrete knowledge elicited from the teacher; instead, the outcome is a multifaceted, complex explanation that combines ideas students already had alongside ideas they negotiated with their teacher and peers in an MBI unit for the purpose of explaining an event that happens in the world. This framing of MBI is more attuned to approximating in classrooms the representation of science as the construction and critique of explanations of events that happen in the world.

While we recognize that the curriculum resources included in the following chapters are necessary supports for shifting instruction in classrooms, we believe the way you think about these supports (i.e., as curriculum resources) will ultimately determine how effective they are in supporting student learning in classrooms, especially in relation to the three-dimensional vision of teaching and learning that prioritizes sensemaking about the world with practices. Given this, we invite you to return to these first two chapters frequently throughout your implementation of the MBI units as you consider how you are thinking about your work with the units in relation to how we think about the work around MBI units, as a way both to collaborate with us, even if asynchronously, and to support continued professional learning through reflection and negotiation. We are excited to share the MBI units with you in the following chapters!

Reference

National Research Council (NRC). 2012. *A framework for K–12 science education: Practices, crosscutting concepts, and core ideas.* Washington, DC: National Academies Press.

SECTION 2 The MBI Units

Unit 1. From Molecules to Organisms

Structures and Processes (LS1): General Sherman

Unit Summary

The General Sherman Tree is an intriguing phenomenon: it is the largest single-stem tree on Earth, yet it grew from a seed that was a mere 4 mm long. In this unit, we engage students in constructing an explanation for how this incredible transformation was possible. To explain this phenomenon, students need an understanding of many of the ideas contained in disciplinary core idea (DCI) Life Science 1 (LS1): From Molecules to Organisms: Structures and Processes, such as the structure-function connections within and between the various levels of organization of an organism (LS1.A), the growth and differentiation of cells as the central mechanism by which organisms increase their size and replace dead or damaged cells (LS1.B), and the fact that cells use processes such as cellular respiration and photosynthesis to obtain energy and matter for their life functions (LS1.C). Taken together, these powerful ideas allow students to construct a scientific explanation of the growth of General Sherman.

Please note that no phenomenon can anchor all the ideas included in a single DCI. This unit does not fully cover ideas related to the ways in which cells sense and respond to stimuli in their environment (LS1.D: Information Processing). We recommend including instruction on this topic after completing the General Sherman unit.

Table 1.1 provides an outline for this unit, describing the purpose and the major tasks and products for each MBI stage, as well as an approximate timeline.

Table 1.1. MBI Unit Outline and Timeline

MBI stage	Purpose	Major tasks and products	Timing
Eliciting ideas about the phenomenon	To introduce the phenomenon, eliciting students' initial ideas to explain the phenomenon, and begin constructing group models	Initial ideas public record; initial group models	1–2 days
Negotiating ideas and evidence through tasks	To provide opportunities for students to make sense of the phenomenon through purposeful tasks and discussions	Individual task products; revised group models	13–16 days
Building consensus	For groups to come to an agreement about the essential aspects of the explanation	Final group models	1 day
Establishing credibility	For each individual student to write an evidence-based explanation	Final evidence-based explanations	1–2 days
		Total	16–21 days

Anchoring Phenomenon

The largest tree in the world is a giant sequoia (*Sequoiadendron giganteum*) in California's Sequoia National Park called General Sherman. General Sherman is tall, standing 274.9 feet (83.8 m) high, and is about 52,500 cubic feet (1,487 m³) in volume—more than half the volume of an Olympic-size pool. Incredibly, this enormous tree started as a small seed, merely 0.16–0.20 inches (4–5 mm) long and 0.04 inch (1 mm) wide.

Driving Question

How did General Sherman gain its mass if it started from such a tiny seed?

Target Explanation

Following is an exemplar final evidence-based explanation that could be expected from students at this grade level by the end of the unit. We have included it to help support you, the teacher, in responsively supporting students in negotiating similar explanations by the end of the unit. It is not something to be shared with students and is provided only as a behind-the-scenes roadmap for you to consider before and throughout the unit to guide your instruction.

General Sherman is the largest tree in the world. A giant sequoia in California's Sequoia National Park, the tree is about 52,500 cubic feet in volume. General Sherman is also tall, standing 274.9 feet high. Incredibly, this enormous tree started as a small seed, merely 4–5 mm long and 1 mm wide. This raises a question: How did General Sherman gain its mass if it started from such a tiny seed?

When the original seed dropped to the forest floor from its parent tree over 2,000 years ago, it became buried in the soil and was exposed to the appropriate environmental conditions (water, oxygen, and proper temperature), sparking germination. In our What's Inside a Seed? task, we dissected seeds and observed the different structures of the seeds, which were also present in General Sherman's seed. The different structures we observed are as follows: the embryo, which develops into the seedling and eventually grows into the roots, stem, and leaves; the cotyledon, which makes up the majority of the seed and provides nutrients to the growing embryo; the seed coat, which acts as a protectant barrier for the seed; and the first leaves.

Initially, General Sherman's seed, which was underground, contained enough stored energy in the form of macromolecules (proteins, lipids, and starch) to start growing. We learned about these macromolecules in our task called What Is the Function of the Endosperm?, in which we tested seeds to see where the biomolecules are found. We found evidence of starch, sugar, and fat but not protein. The energy General Sherman used to grow came from the process of cellular respiration ($C_6H_{12}O_6 + 6O_2 \rightarrow 6CO_2 + 6H_2O + ATP$). During cellular respiration, the cells' mitochondria convert chemical energy stored in carbohydrates (glucose) into usable energy for the cells in the form of ATP. Cells use ATP for various processes, including mitosis, which allows the tree to grow. We learned about cellular respiration during our How Do Trees Use Sugar? task, in which we tested germinating seeds for cellular respiration. We found that they take in oxygen and give off carbon dioxide just like animals do. Also, in our task called How Does the Endosperm Produce Sugar?, our tests showed that leaves produce carbohydrates in the light but not in the dark and that the leaves take in CO_2 and give off O_2 during respiration.

As General Sherman grew, it gained mass. In order to grow, it needed to add many trillions of cells to its body through the process of mitosis. During mitosis, a cell replicates its genetic material along with cellular organelles before dividing into two daughter cells. In our How Do Plants Grow? task, we learned that most cells look the same except when they are in the process of dividing. During this process, the nucleus looks different. As General Sherman used ATP to carry out the process of mitosis, the number of cells increased; consequently, the mass of the tree also increased.

The majority of a tree's mass comes from cellulose ($C_6H_{10}O_5$), the carbohydrate that makes up the cell walls of plants. We saw this in the Where Does All That Matter Come From? task, which showed that the water was mainly lost and the increase in plant mass was more than the mass of the soil that was lost. Cellulose is a polymer of glucose molecules ($C_6H_{12}O_6$). These cell walls provide support and structure to each plant cell. In our What's Inside a Tree? task, we learned about cell differentiation in trees and that one of the most abundant types of cells in any tree is xylem cells. The wood that makes up the trunk and branches of the tree is composed of these xylem cells (tiny tubes that transport water and dissolved nutrients). From its roots, General Sherman obtained essential nutrients that the tree needs to function.

As General Sherman produced more and more needles, small openings in the leaves called stomata allowed the transfer of air in and out of the needles. The carbon dioxide from the air diffuses into the cells and is fixed using energy from the sun during the process of photosynthesis ($6CO_2 + 6H_2O \rightarrow C_6H_{12}O_6 + 6O_2$). In our What's Inside a Tree? task, we also learned that the part of the cell that is responsible for this process is the chloroplast, a cell organelle that gives needles and leaves their green color. In addition to the carbon dioxide taken in from the air, water is also required for photosynthesis to take place. This water is taken up by the roots of the tree and transported throughout the tree to the leaves. Glucose is also a byproduct of photosynthesis and can be used as a source of energy during cellular respiration or can be stored as larger polysaccharides such as cellulose or starch.

Where General Sherman is growing is important as well. In the task called Does the Environment Affect Plant Growth?, we saw that photosynthesis works best with bright light and warm temperatures. We looked at tree rings and found that the trees had not grown as much during droughts or when it was especially cold. In our final task, How Do Trees Survive in Different Climates?, we looked at the climatic conditions of Sequoia National Park, which has cold, wet winters and hot, dry summers. We saw that the leaves, branches, and roots are adapted to help photosynthesis by reducing water loss when it is hot and preventing freezing when it is cold. These adaptations have helped General Sherman grow for thousands of years!

Unlike humans, General Sherman, like all plants, must make its own food. Therefore, trees must get their mass from water, soil, sunlight, or air. The majority of General Sherman's mass came from CO_2 that was converted into glucose ($C_6H_{12}O_6$) molecules, the major product of photosynthesis.

Eliciting Ideas About the Phenomenon

The eliciting ideas about the phenomenon stage of MBI is about introducing the phenomenon and driving question, eliciting students' initial ideas about what might explain the phenomenon and answer the driving question, and constructing their initial models in small groups. We have provided a PowerPoint that will lead you step-by-step through the process, which can be downloaded from the book's Extras page at *www.nsta.org/mbi-biology*. There are two products of this stage: an Initial Hypotheses List and student groups' initial models. Examples of each are shown in Figures 1.1 and 1.2.

This stage of MBI is critically important, as it orients students to the MBI unit anchoring phenomenon using the driving question in a

Figure 1.1. Example Initial Hypotheses List

Initial Hypotheses

- Perfect conditions → sunlight, dirt, water
- Source of mass must be in addition to the seed
- Nutrients from water, energy from sunlight
- General Sherman had no competition for resources → lots of growth
- Resources were unlimited causing growth
- Very long period of time to grow, still growing
- Relatively stable environmental conditions
- Grew around another tree (parasitism)
- Nothing stopping growth (humans, other organisms, etc.)
- Genetically predisposed to be very large

Figure 1.2. Example Initial Model

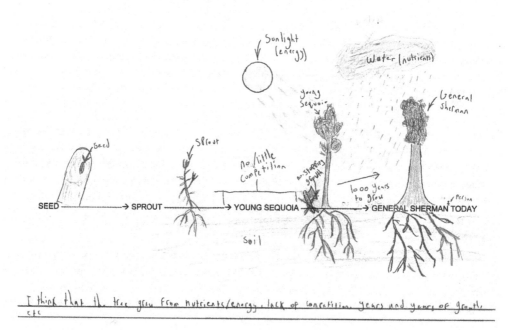

General Sherman Initial Model

How did General Sherman gain its mass if it started from such a tiny seed?

way that makes it compelling enough to spend several weeks across a unit working on explaining an event. Put more succinctly, this stage introduces the problem space in which the students will be working. Introducing the phenomenon should be done in a way that supports students in drawing on what they have previously learned in and outside of school as they begin to think about how to explain the driving question. Your priority should be to elicit students' ideas about the phenomenon, since it is important for them to continually think about the phenomenon and refine their everyday ways of thinking. Since eliciting students' ideas is a top priority, it's important to create a learning environment where students feel comfortable and are invited to offer ideas. This means you need to think carefully about how to get students to float (or put on the table for consideration) as many ideas as possible. At this stage, there are no right or wrong ideas. Everything (within reason!) is on the table.

Negotiating Ideas and Evidence Through Tasks

The negotiating ideas and evidence through tasks stage of MBI makes up the majority of the unit. Each task is designed to introduce a key scientific idea or to reinforce an idea already raised by students in the class. Halfway through the unit, we recommend having groups revise their models and share out those revisions. PowerPoint templates are provided for the tasks as well as for the model revision; these can be downloaded from the book's Extras page at *www.nsta.org/mbi-biology*. Each task begins by introducing ideas for students to reason with while working on the task. During the task, students investigate or test the concepts using data. We recommend using the Back Pocket Questions provided in the Teacher Notes to press students' thinking while they are engaged in the task. Taken together, the tasks scaffold students' thinking as they co-construct a scientific explanation of the phenomenon. The general outline of this stage is as follows:

Task 1. What's Inside a Seed?

Task 2. What Is the Function of the Endosperm?

Task 3. How Do Trees Use Sugar?

Task 4. How Does the Endosperm Produce Sugar?

Midunit Model Revision for Unit 1

Task 5. How Do Plants Grow?

Task 6. Where Does All That Matter Come From?

Task 7. What's Inside a Tree?

Task 8. Does the Environment Affect Plant Growth?

Task 9. How Do Trees Survive in Different Climates?

A Summary Table is used after each task to scaffold students' negotiation of ideas and how those ideas help explain the phenomenon. An example is provided in Table 1.2. Like the target explanation, this is included here to serve as a behind-the-scenes roadmap to help support you, the teacher, in responsively supporting students in negotiating similar responses by the end of the unit. It is not something to be shared with students but is provided for you to consider before each task to guide your instruction.

Table 1.2. Example Summary Table for Unit 1

Task	What we learned from this task	How it helps us explain the anchoring phenomenon
1. What's Inside a seed?	All seeds have similar structures inside. All seeds contain an embryo, a seed coat, and endosperm. Seeds begin to grow (germinate) when water is added to them. The first structure to appear is roots. The next structures are a stem and small leaves.	General Sherman only needed water to start growing. The embryo inside a tree seed increases in size once it germinates and becomes a seedling, and the endosperm gets smaller.
2. What is the Function of the Endosperm?	The endosperm is made of starch, sugars, and fats. Starch, sugars, and fats are molecules that store chemical energy.	The endosperm could be an energy source for the embryo as it germinates because it is made up of starch and sugars.
3. How Do Trees Use Sugar?	Cellular respiration is a chemical process in which the bonds of sugar molecules and oxygen molecules are broken and new compounds are formed that can transport energy to cells and tissues throughout the plant.	The embryo will use the starch and sugar in the endosperm for cellular respiration. Once the endosperm is used up, there is not a source of sugar for cellular respiration.
4. How Does the Endosperm Produce Sugar?	The main way that solar energy is captured and stored on Earth is through a chemical process known as photosynthesis. This process converts light energy to stored chemical energy by converting carbon dioxide plus water into sugars plus released oxygen.	General Sherman is able to convert light energy to stored chemical energy by converting carbon dioxide plus water into sugars plus released oxygen. These sugars are then used for cellular respiration. General Sherman is able to produce all the sugar it needs to keep growing.
5. How Do Plants Grow?	In plants, like all multicellular organisms, individual cells grow and then divide via a process called mitosis, thereby allowing the organism to grow.	The cells inside of General Sherman increase in number as the tree grows. The cells do not simply grow in size. Each new cell that is created has the same DNA as the original cell so the whole plant has the information it needs to function and keep growing.
6. Where Does All That Matter Come From?	Most of the matter that makes up a plant comes from the air. The sugar molecules formed through photosynthesis contain carbon, hydrogen, and oxygen. The hydrocarbon backbones of sugars are used to make amino acids and other carbon-based molecules that can be assembled into larger molecules (such as cellulose, proteins, or DNA) and then used to create new cells.	The sugar that General Sherman makes through the process of photosynthesis can also be used to make amino acids and other carbon-based molecules. The carbon-based molecules can then be combined into even larger molecules such as cellulose, proteins, and DNA. The larger molecules are used to form new cells, which are combined into tissues. Tissues are then organized into systems that carry out the functions of the tree.

Continued

Table 1.2. Example Summary Table for Unit 1 *(continued)*

Task	What we learned from this task	How it helps us explain the anchoring phenomenon
7. What's Inside a Tree?	Photosynthesis mainly takes place in the leaves of trees. Leaves are made up of specialized cells that help them perform the essential functions of life. Plants have vascular tissues that transport water from the ground to the leaves and sugar and other carbon-based molecules from the leaves to the rest of the plant. Leaves are also needed to move water through a plant.	General Sherman gets the carbon dioxide it needs for photosynthesis through pores in its needles called stomata. Water enters the tree through the roots, moves up the trunk through the xylem, and then enters the needles. It also exits the leaves through the stomata.
8. Does the Environment Affect Plant Growth?	Photosynthesis is most efficient with brighter light and warmer temperatures. If it gets too hot or too cold, chlorophyll breaks down and photosynthesis stops. Trees do not grow as much during droughts or colder weather.	General Sherman needs constant access to bright light, water, and warm temperatures in order to grow. When it gets too cold or there is a drought, it does not grow as well.
9. How Do Trees Survive in Different Climates?	The leaves, branches, and roots of plants are adapted to maximize photosynthesis and minimize water loss in hot climates and damage from water freezing in cold climates. Sequoia National Park has cold, wet winters and hot, dry summers. The area gets about 45 inches of rain annually, and snow is common during the winter months.	The leaves, branches, and roots General Sherman are adapted to (a) minimize water loss, (b) maximize water absorption, and (c) prevent cell damage during freezing temperatures. These adaptations helped this tree survive and grow for thousands of years.

You can think of your role here as that of helping introduce ideas to students to support them in learning about and beginning to engage with the ideas through carefully designed sensemaking tasks. If important ideas are not introduced by the students, you can put them on the table or introduce them through the use of just-in-time direct instruction, short videos, or readings, among other strategies. Since the objective is to have students pick up and try out these ideas in carefully crafted tasks as they develop explanations, introducing your students to an idea early in the task is *not* giving them answers that will be confirmed in the task. Instead, you can think of this as introducing students to tools (i.e., disciplinary core ideas) that will be useful in a task in ways that will help them make sense of data, a simulation, an activity, or something in the world. In the end, it is important that you think of introducing new ideas in the early parts of this MBI stage as an opportunity for students to engage in sensemaking with new ideas to use, while you concurrently recognize the potential explanatory power of the new ideas. At the end of each task, the completion of a row of the Summary Table is an opportunity for students to think about what they have learned in the task and, together with peers, reason about how the newly introduced ideas can be applied to support their objective of explaining the unit anchoring phenomenon.

Building Consensus

In the building consensus stage of an MBI unit, students work to build a class consensus about the explanation of the phenomenon by finalizing the groups' models, comparing and contrasting those models as a whole class, and constructing a consensus checklist of the ideas and evidence (called the Gotta Have Checklist) that should be included in their final evidence-based explanations that make up the summative assessment of the unit. We have provided a PowerPoint template as a guide through this stage of the unit, which can be downloaded from the book's Extras page at *www. nsta.org/mbi–biology*. Examples of the whole-class Gotta Have Checklist and a group's final model appear in Figures 1.3 and 1.4.

Figure 1.3. Example Gotta Have Checklist

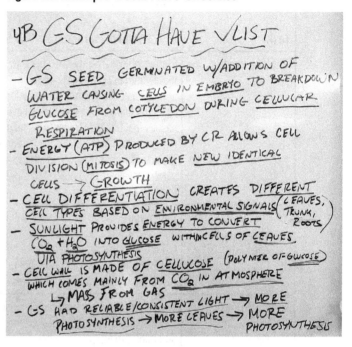

In Chapter 2, we described the scientific practice of modeling as a knowledge-building practice and models as the products of modeling. Students begin this stage by finalizing their models in groups of three or four, which they should think of as a continuation of the sensemaking that started in the groups' initial modeling experiences in the early days of the MBI unit and continued when they revisited these models midway through the unit. As with these other experiences, you can expect the practice of modeling to support groups in refining their explanations of the unit anchoring phenomenon. Here, negotiation and argumentation lead to refined ways of thinking about the phenomenon that may not have occurred had students not engaged in modeling this final time. The groups' models will offer insight into your students' thinking that otherwise may not be accessible to you. At the same time, models as products of modeling also serve as artifacts that students can compare across groups in ways that are supported by your selection and sequencing of groups' sharing in whole-class sensemaking, so differences can be foregrounded, negotiated, and resolved through argumentation, among other science practices.

Finally, the construction of final criteria or whole-class mapping of ideas included in the Gotta Have Checklist to the evidence collected from the various tasks completed in the second stage of MBI is a scaffold that will be important for individual students in writing evidence-based explanations. Importantly, the teacher-facilitated, co-constructed Gotta Have Checklist outlines what the class agrees should be included in students' explanations and is mapped to evidence identified across the unit tasks. This step is the last stage of building consensus. This is especially true because final group models will have been negotiated in both small and whole groups as a basis for the identification of the criteria that are needed for writing an evidence-based explanation.

Figure 1.4. Example Final Model

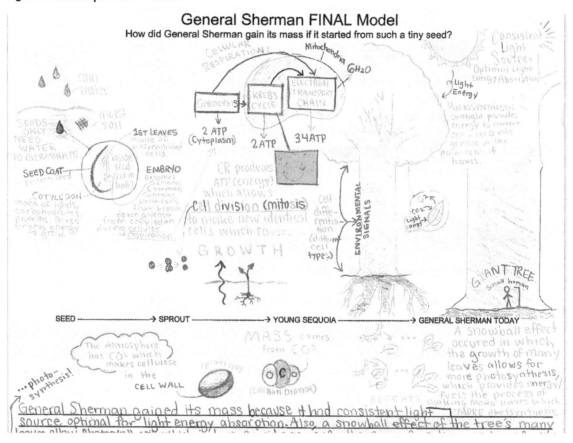

Establishing Credibility

In MBI, students must argue for their ideas in writing. In the establishing credibility stage, they do this through written evidence-based explanations, peer review, and revision. We have provided a PowerPoint template as a guide through this stage of the unit, which can be downloaded from the book's Extras page at *www.nsta.org/mbi-biology*. Writing scaffolds such as sentence stems may be necessary to jump-start some students' writing. Once the written explanations are complete, we recommend using the MBI Explanation Peer-Review Guide found in Appendix B. The target explanation on pages 22–23 is an example.

Writing evidence-based explanations is a sensemaking experience that students engage in early in this stage of MBI. This is another knowledge-building opportunity that, like engaging in modeling earlier in the unit, is a practice that leads to a product. And like models, which served as products that could provide you with insight into the way groups were thinking about explaining the anchoring phenomenon, the individually written evidence-based explanations afford you with an opportunity to assess individual students in terms of where they are in their attempts to explain the anchoring phenomenon. As students engage with others in this stage, this is an additional space for argumentation and negotiation as they recognize possible differences and details they may not have considered or may not yet agree with.

This practice of peer review needs to be supported, so we have provided the MBI Explanation Peer-Review Guide in Appendix B, which students can use to provide feedback to each other. The peer-review guide contains questions designed to focus student attention on the most important aspects of explanations and to encourage students not only to discuss what they figured out and how they know what they know but also to best communicate what they have learned to others in a way that is clear, complete, and persuasive. During the peer review, you should consider and make explicit how explanations, like models, are refined over time through negotiation within a community. The MBI unit culminates at the end of this stage as students use the feedback they have received to make final revisions that represent how they are thinking about the unit anchoring phenomenon. The final evidence-based explanations serve as one measure of what students have learned about engaging in science practices to use disciplinary core ideas and crosscutting concepts to explain the phenomenon.

Hints for Implementing the Unit

- One challenge students often have with this unit is moving from a macro to a micro focus. While a full explanation of the growth of General Sherman does require a macro perspective at times, helping students "see" what is happening at the micro level is essential to this unit. We recommend supporting students in shifting to a micro lens through questioning during tasks and whole-group discussions. When modeling, you might suggest that they include a zoom-in window to show what is happening at the molecular level inside a leaf undergoing photosynthesis, for example.

- We recommend providing opportunities to connect students' personal experiences and community resources with the phenomenon and topic of this unit. While students may not have had any personal experiences with the phenomenon, we suggest holding discussions throughout the unit to elicit their personal connections. Think about possible connections in your community that you can bring to the conversation as well. These kinds of connections will make the unit personally meaningful to your students, increasing their motivation to engage in the tasks. In our experience, much of this happens during the eliciting ideas about the phenomenon stage of the unit. However, you should prompt students to connect to the topic during all phases of the unit.

- There may be times during the unit when students make connections to injustices or disparities in their communities or raise emotional responses to the topics. We suggest preparing to address young people's questions and desire for activism so science practices are not portrayed as disconnected from the social and cultural contexts of students' real-world experiences.

*Targeted **NGSS** Performance Expectations*

- **HS-LS1-1.** Construct an explanation based on evidence for how the structure of DNA determines the structure of proteins which carry out the essential functions of life through systems of specialized cells.

- **HS-LS1-2.** Develop and use a model to illustrate the hierarchical organization of interacting systems that provide specific functions within multicellular organisms.

- **HS-LS1-3.** Plan and conduct an investigation to provide evidence that feedback mechanisms maintain homeostasis.

- **HS-LS1-4.** Use a model to illustrate the role of cellular division (mitosis) and differentiation in producing and maintaining complex organisms.

- **HS-LS1-5.** Use a model to illustrate how photosynthesis transforms light energy into stored chemical energy.

- **HS-LS1-6.** Construct and revise an explanation based on evidence for how carbon, hydrogen, and oxygen from sugar molecules may combine with other elements to form amino acids and/or other large carbon-based molecules.

- **HS-LS1-7.** Use a model to illustrate that cellular respiration is a chemical process whereby the bonds of food molecules and oxygen molecules are broken and the bonds in new compounds are formed resulting in a net transfer of energy.

Eliciting Ideas About the Phenomenon Stage Summary

The first stage of MBI, eliciting ideas about the phenomenon (Figure 1.5), involves introducing the anchoring phenomenon and driving question, eliciting students' initial ideas and experiences that may help them develop initial explanations of the phenomenon, and developing initial models of the phenomenon based on those current ideas. Before beginning this stage, review the relevant sections in Chapters 1 and 2 for specifics. See Figure 1.1 on page 24 for an example of an Initial Hypotheses List for this unit. We have provided a PowerPoint template to assist with this stage, which can be downloaded from the book's Extras page at *www.nsta.org/mbi-biology*.

Figure 1.5. Eliciting Ideas About the Phenomenon Stage Summary

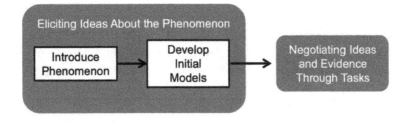

Introduce the Phenomenon

We begin this stage by introducing the phenomenon in an engaging way, such as with stories, videos, demonstrations, or even short activities. The goal is to provide just enough information for students to begin to reason about the phenomenon, without providing too much of the explanation. While we are introducing the phenomenon, we ask questions to keep students engaged and make sure they are paying attention to the important aspects that can begin to help them explain the phenomenon. The introduction ends with the driving question of the unit, which we have found helps focus their thinking on the development of a causal explanation for the phenomenon, or a "why" answer.

It is important to spend time eliciting not only students' scientific ideas about the phenomenon but also their personal connections to the phenomenon. What personal experiences do they have that help them connect to the phenomenon? While the phenomenon of the unit may not be directly connected to your community, does the community have similar phenomena or specific resources that you can bring into the discussion? The more connected students feel to the unit phenomenon, the more engaged and motivated they will be.

We then get students into their groups and facilitate the first discussion to try to answer the driving question with just the resources they brought with them—the ideas, experiences, and cultural resources they have gained both in and outside of the classroom. These ideas may be fully formed, partially correct, or fully incorrect in terms of our canonical knowledge of science. However, we make it clear that all ideas are considered equally valid at this point in time, as we realize that these are the ideas students put into play when they think about the phenomenon we have introduced.

Once student groups have discussed their ideas, we facilitate a class discussion to compare and contrast the ideas generated by each group. As ideas are presented, they are put on our first public record, which we call the Initial Hypotheses List. We use this list throughout the unit to keep track of the changes in students' thinking as they work toward a final evidence-based explanation for the anchoring phenomenon. We consider all ideas to be valid at this stage, before students use other resources to begin making sense of how or why something happens. The Initial Hypotheses List is also useful for the next task in this stage, initial model construction, especially since it offers students additional ideas beyond those they initially had either individually or in their small groups.

Develop Initial Models

If students are not experienced with modeling, it is worth providing a brief introduction and example. We have included an example in the PowerPoint for this stage, which can be downloaded from the book's Extras page at *www.nsta.org/mbi-biology*. Once the class is ready to begin modeling, we give each group a sheet of 11 × 17 inch paper and ask the groups to each make a model of their initial hypothesis. Sometimes they choose their own original hypotheses, while other times they are influenced by their peers' ideas and adopt one of them instead. As the groups work on constructing their models, you should walk around asking clarifying questions and pushing them to be as specific as possible. Once the models are ready, it is important to have students share ideas across them. There are a number of ways you might run these share-out sessions. We often collect and present the models on a document camera at the end of the first day. Groups can provide one- or two-sentence summaries of the initial hypotheses that they have represented in their models. We point out interesting ideas and ways in which they have represented these ideas. For example, we may call attention to the fact that a group labeled the arrows, which made the model more understandable, and that another group used a zoom-in window to show what was happening at a different scale. At the end of the first day and this first stage, we have elicited ideas across the class, and the groups' initial hypotheses and models will act as a starting point for the rest of the unit.

Negotiating Ideas and Evidence Through Tasks Stage Summary

Figure 1.6. Negotiating Ideas and Evidence Through Tasks Stage Summary

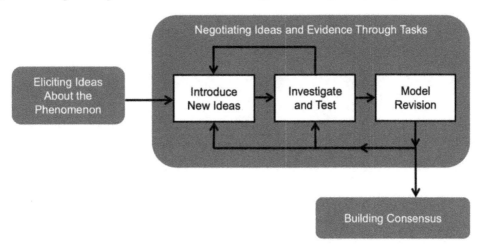

The next stage of MBI, negotiating ideas and evidence through tasks (Figure 1.6), takes up the majority of the unit. The tasks are designed to introduce or extend important science ideas that students need as they construct their evidence-based explanations of the phenomenon. Before beginning the tasks, we suggest that you review the relevant sections in Chapters 1 and 2 for specifics. Each task consists of the following:

- A **Teacher Notes** section that provides an overview of the activity and science content important to the task. Here, we provide guidance on conducting the task with details on the procedure, a list of required materials and preparation, any necessary safety precautions, suggested Back Pocket Questions, and an example Summary Table entry. At the end of this section are further hints for implementing the task and possible extensions, if desired.

- A **Student Handout** to be given to groups as they engage in the task. The handout provides an introduction to the task and content, an Initial Ideas section to frame the learning before they begin, and detailed instructions and work space. It ends with a section called What We Figured Out, which is designed to scaffold student responses to include in the Summary

Table as a whole class. We suggest having students fill out this section after completing the task and the post-task discussion.

- A **PowerPoint** Negotiating Ideas and Evidence Through Tasks template you can adapt for each task, which can be downloaded from the book's Extras page at *www.nsta.org/mbi-biology*.

- A Midunit Model Revision reminder roughly halfway through the unit.

Teacher Notes

Task 1. What's Inside a Seed?

Purpose

This task gives students an opportunity to learn more about the internal structure of a seed and what happens to the parts of the seed when it germinates. To accomplish this task, students dissect beans, seeds, and nuts and then watch several videos that show what happens to a seed when it germinates. These ideas help students understand how the life cycle of large multicellular plants, such as General Sherman, began as a small seed.

Important Life Science Content

The three main parts of a seed are the embryo, endosperm, and seed coat. The embryo is the most important part of the seed because it contains all the cells necessary to mature into a developed plant. The embryo itself has three main parts: the primary root (radicle), endosperm, and embryonic leaves. Once germination occurs, the primary root is the first part to appear from the seed. The primary root provides a stable support system for the growing seedling. The endosperm supplies nutrients for the entire embryo during the process of germination. The embryonic leaves are the first leaves to emerge aboveground and are accompanied by the endosperm in order to receive all available nutrients until they are depleted. The seed coat, which can be thick or thin, is the outermost part of the seed and protects all the internal parts of the seed by preventing physical, temperature-related, or water damage. The seed coat ensures that the internal parts remain inactive until optimal conditions are met for germination.

Scientific Ideas That Are Important to Think About During This Task

- All seeds have similar structures inside. All seeds contain an embryo, a seed coat, and endosperm.
- Seeds begin to germinate (grow or develop) when water is added to them. The first structure to appear is the roots. The next structures are a stem and small leaves.

Timeline

Approximately one class period.

Materials and Preparation

The items needed for this investigation are listed in Table 1.3 (p. 36). You may purchase the seeds from a big-box retail store such as Walmart or Target or through an online retailer such as Amazon. Be sure to buy seeds intended for planting and not ones meant for consumption. The other materials may be purchased through a science education supply company such as Ward's Science, Flinn Scientific, or Carolina Biological Supply.

Table 1.3. Required Materials for Task 1

Item	Quantity
Safety goggles, nonlatex apron, and vinyl or nitrile gloves	1 per student
Scalpel	2 per group
Hand lens	2 per group
Forceps	2 per group
Petri dish	1 per student
Electronic scale	1 per group
Pinto bean seeds	4 per group
Lima bean seeds	4 per group
Kidney bean seeds	4 per group
Corn seeds	4 per group
Pine nuts (pine seeds)	4 per group
Sunflower seeds	4 per group
Pea seeds	4 per group
Computer or tablet with internet access	1 per group
500-ml beakers	7 per class
Paper towels	1 roll per class
Student Handout	1 per student

This task has students explore seed germination using the different kinds of seeds in the materials list. Label one beaker for each type of seed, and then place the seeds in the appropriate beakers. Fill with water and soak the seeds for at least 24 hours before beginning this task. Before class begins, simply drain off the water. Place the materials in a central location in the classroom so students can get them during the task.

Safety Precautions

Remind students to follow all normal safety rules. In addition, tell students to take the following safety precautions:

- Wear sanitized indirectly vented chemical-splash goggles, nonlatex aprons, and vinyl or nitrile gloves during setup, hands-on activity, and cleanup.

- Be careful with scalpels. They are sharp and can cut or puncture skin.

- Use caution handling glass labware, which can shatter and cut skin or eyes.
- Never taste or eat any food items used in a lab activity.
- Clean up any spills on the floor immediately to avoid a slip or fall hazard.
- Wash hands with soap and water when the activity is completed.

Procedure

This lesson plan is only a suggestion. It is included here to illustrate how you can facilitate student thinking during this task. We encourage you to modify this lesson plan by asking different questions, using different examples, and providing different scaffolds as appropriate to better meet the needs of students in your class.

Introduction of the Task (10 minutes)

1. Have students sit in groups of three or four.

2. Give each student one copy of the Student Handout for Task 1.

3. Read the Introduction and Initial Ideas sections of the handout aloud, having students follow along. It is a good idea to show a picture of General Sherman and seeds from a giant sequoia at this point in the lesson so students have the anchoring phenomenon of the unit in mind as they start this task.

4. Ask students to draw the structures that they think are inside a seed on their own on their handouts. Remind them to be sure to describe the functions of those structures in their pictures.

5. Give students an opportunity to share their pictures with the others in their group.

6. Ask one student from each group to share how their ideas about what is inside a seed and the functions of these structures were similar to or different from the ideas of others in the group.

Making Observations and Taking Measurements (30 minutes)

1. Read the section of the handout called Your Task aloud to students.

2. Show them the available materials and ask how they might use some of the equipment, such as the scalpel, petri dish, and electronic scale. Don't be afraid to offer some suggestions, but do not require that every group do everything the same way. Give them a choice.

3. Go over the safety precautions for this task.

4. Give students about 15 minutes to dissect the seeds. Have them record what they observe or measure on their handouts.

5. As students work, move from group to group and check in with them. It's important to ask them questions that will help them connect what they are doing to the goal of the task and the anchoring phenomenon. (See the Back Pocket Questions for Task 1 on pages 38–39 for some suggestions.)

6. Ask students to clean up their workstations and return the materials. Be sure to keep the dissected seeds. **Students will use the dissected parts in Task 2.**

7. Give students about 10 minutes to watch the three videos of seeds germinating. Remind them to record what they observe on their handouts.

Putting Ideas on the Table (10 minutes)

1. Give an interactive lecture to introduce some ideas that students might find useful for making sense of what they did. We recommend, as a minimum, introducing the following ideas:

   - Names of the three main parts of seeds: the embryo, endosperm, and seed coat (but *not* the functions of these structures).

   - Names of different parts of seedlings, such as roots, stem, and leaves (but *not* the functions of these structures).

   - The idea that trees are multicellular organisms, which, like all multicellular organisms, are a system of interacting subsystems. These subsystems are groups of cells that work together and are specialized for particular functions.

2. Encourage students to keep a record of the ideas they find important in the section called Some Useful Ideas From My Teacher on the Student Handout so they can refer to them later in the unit. Remember, your goal here is to put some ideas on the table to help your students make sense of what they are seeing and doing, not to tell them what they should have learned.

Adding Information to the Summary Table (10 minutes)

1. Give students 5 minutes to decide what to add to the Summary Table at the end of the Student Handout.

2. Have one student from each group share what the group figured out, how they know (their evidence for what they figured out), and how this information will help them explain the anchoring phenomenon.

3. Once each group has shared, ask the entire class to decide what should be added to each column of the class Summary Table for Task 1. Help students reach consensus about what to add to the Summary Table. Only add an idea to the Summary Table if everyone in the class agrees with that idea.

Back Pocket Questions

As students work in groups, it is important to engage with each group to help press and extend students' thinking around the ideas at play in this task. Following are some example questions you might ask:

1. Helping students get started: What parts of the seed are you seeing? What do you think they do?

2. Pressing further: This seed will eventually be a larger plant. How do you think the parts of the seed you're seeing become a large plant?

3. Following up: What makes you think that? Can you say more?

Filling Out the Summary Table

Table 1.4 includes examples of the responses students may come up with when they fill out the Summary Table. This is provided here only as a behind-the-scenes roadmap and is not meant to be shared with students.

Table 1.4. Example Summary Table for Task I

What we learned	How it helps us explain the phenomenon
All seeds have similar structures inside. All seeds contain an embryo, seed coat, and endosperm. Seeds begin to grow (germinate) when water is added to them. The first structure to appear is roots. The next structures are a stem and small leaves.	General Sherman only needed water to start growing. The embryo inside a tree seed increases in size once it germinates and becomes a seedling, and the endosperm gets smaller.

Hints for Implementing This Task

- As mentioned earlier, it is important that you buy seeds meant for planting and not consumption!

- Students may have experience with germinating seeds while gardening at home. This is a great place for a student with specialized knowledge to shine.

Possible Extensions

- Additional seeds, including large tropical seeds, can be used to show diversity in seed structure. Coconuts are the largest seed in the world!

Student Handout

TASK 1. WHAT'S INSIDE A SEED?

Introduction

General Sherman is the largest tree in the world. A tree, like all multicellular organisms, is a system of interacting subsystems. These subsystems are groups of cells that work together and are specialized for a particular function. One such subsystem is the reproductive system. The function of the reproductive system of trees, as in most plants, is to produce seeds. To explain how General Sherman got so big, you will therefore need to first figure out what is inside a seed and what happens to the parts of a seed when it germinates. This is important because the function of a living thing depends on the properties of its parts as well as the relationships among those parts.

Initial Ideas

Before beginning this task, take a few minutes to think about the structures you might find inside a seed and what the functions of those structures might be. In the space below, draw and label the structures you think are found inside a seed. Be sure to include a description of what you think each structure does in your labels.

Your Task

Your task is to use what you know about the relationship between *structure and function* in living things to *plan and carry out an investigation* to determine what is inside a seed and what happens to the parts of a seed when it germinates. To accomplish this task, you will cut open and examine the contents of at least six different kinds of seeds. You will then watch a time-lapse video of a germinating seed. Be sure to keep track of anything you observe or measure as you work in the Observations and Measurements section of your handout.

Available Materials

You and your group may use any of the following materials during this task:

- Safety goggles, nonlatex apron, and vinyl or nitrile gloves (required)
- Scalpel
- Hand lens
- Forceps
- Petri dish
- Computer or tablet with internet access
- Electronic scale

- Pinto bean seeds
- Lima bean seeds
- Kidney bean seeds
- Beakers
- Corn seeds
- Pine nuts (pine seeds)
- Sunflower seeds
- Pea seeds
- Paper towels

Safety Precautions

Follow all normal lab safety rules. In addition, be sure to take the following safety precautions:

- Wear sanitized indirectly vented chemical-splash goggles, nonlatex aprons, and vinyl or nitrile gloves during setup, hands-on activity, and cleanup.
- Be careful with scalpels. They are sharp and can cut or puncture skin.
- Use caution handling glass labware, which can shatter and cut skin or eyes.
- Never taste or eat any food items used in a lab activity.
- Clean up any spills on the floor immediately to avoid a slip or fall hazard.
- Wash hands with soap and water when the activity is completed.

Observations and Measurements

Use the space below to keep a record of what you observe or measure as you work with the seeds.

Activity

Watch the following videos to see what happens to different types of seeds after they germinate and begin to grow:

- *www.youtube.com/watch?v=sMK-BKUYM0s&t=16s*
- *www.youtube.com/watch?v=w77zPAtVTuI&t=7s*
- *www.youtube.com/watch?v=Y6vgAnMhGxs&t=6s*

Keep a record of what you notice as you watch these videos in the space below.

Some Useful Ideas From My Teacher

You can keep track of useful ideas from your teacher in the space below.

What We Figured Out

Now that you have completed this task, take a few minutes to fill out the Summary Table below with the other students in your group. This table will help you keep track of what you figured out during the task. You will then have an opportunity to share your ideas as we fill out our class Summary Table.

Summary Table

What we learned	How we know

How it helps us explain the phenomenon

Teacher Notes

Task 2. What Is the Function of the Endosperm?

Purpose

This task gives students an opportunity to learn more about the biomolecular composition of the endosperm in order to make inferences about its function. To accomplish this task, students use indicators to test for the presence of carbohydrates, lipids, and proteins in the endosperm. This information helps students understand why the embryo in a seed gets bigger over time after germination but the endosperm gets smaller.

Important Life Science Content

Plants, like all living things, are made up of large, complex molecules called biomolecules. There are four main types of biomolecules: carbohydrates, lipids, proteins, and nucleic acids.

Carbohydrates contain the elements carbon, hydrogen, and oxygen in a specific ratio represented by the formula CH_2O. There are many different molecules that are classified as carbohydrates. Some are small molecules, such as the sugars glucose and sucrose (Figure 1.7); others are large molecules, such as cellulose and starch (Figure 1.8). Living things use carbohydrates to store chemical energy. This chemical energy is stored in the bonds that hold the atoms that make up the molecule together. More energy can be stored in larger carbohydrates, such as starch, because these molecules have more atoms and thus more bonds holding them together than do smaller molecules, such as glucose or sucrose (see Figures 1.7 and 1.8). Plants also make structures such as cell walls out of carbohydrates.

Figure 1.7. Glucose and Sucrose Molecules

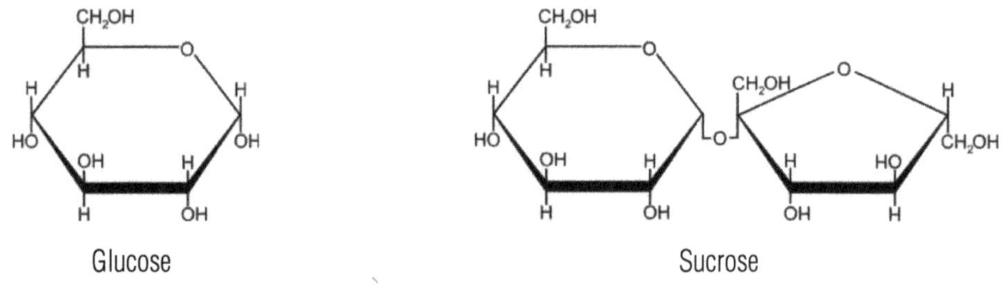

Glucose   Sucrose

Figure 1.8. Starch Molecule

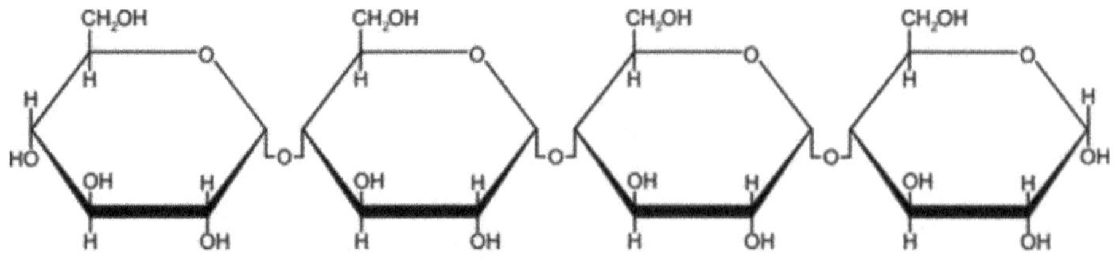

TASK 2. WHAT IS THE FUNCTION OF THE ENDOSPERM?

Benedict's solution can be used to test the presence of glucose in the endosperm. The blue solution turns green with trace amounts of sugar, orange with moderate amounts, and red with large amounts. Lugol's iodine solution can be used to test the presence of starch in the endosperm. The reddish-yellow solution turns black in the presence of a starch but stays the same color when no starch is present.

Lipids do not share a common molecular structure like carbohydrates. The most common lipids are called triglycerides, which include fats and oils. Triglycerides have a glycerol backbone bonded to three fatty acids (Figure 1.9). Waxes and steroids are other kinds of lipids. Living things use lipids to store chemical energy and also create structures such cell membranes out of lipids. A piece of brown paper, such as a grocery bag, can be used to test for the presence of lipids in the endosperm. Lipids create a grease spot on the brown paper.

Figure 1.9. Example of a Lipid

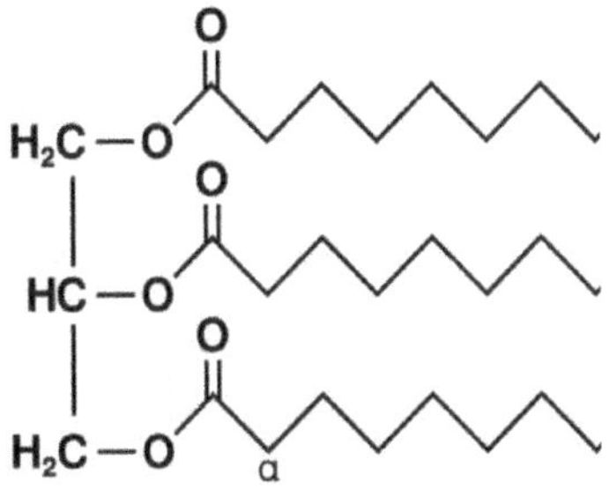

Proteins are large, complex molecules that carry out most functions of the cells and tissues found in plants. They are responsible for helping other molecules get in and out of cells, speeding up chemical reactions, producing other molecules, and breaking down waste. Proteins are made up of hundreds or thousands of smaller units called amino acids. There are 20 different types of amino acids (Figure 1.10), which are made up of carbon, hydrogen, oxygen, nitrogen, and sulfur atoms. The sequence of amino acids in a protein determines the function of the protein. Biuret solution can be used to test for the presence of proteins in the endosperm. The solution turns purple, violet, or pink in the presence of a protein (or amino acids) but stays blue when no protein is present.

Figure 1.10. Example of an Amino Acid Found in Protein

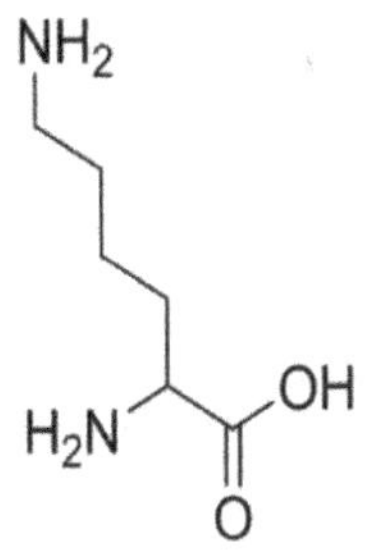

Nucleic acids are made up of carbon, hydrogen, oxygen, nitrogen, and phosphorus atoms. Plants, like animals, use nucleic acids to store or transfer information about how and when to make the many proteins needed to build and maintain functioning cells, tissues, and organisms. They also use nucleic acids to create new proteins, move amino acids around inside cells so they can be used to make new proteins, and regulate the process of creating proteins.

Scientific Ideas That Are Important to Think About During This Task

- The endosperm is made of starch, sugars, and lipids.
- Starch, sugars, and lipids are molecules that store chemical energy.

Timeline

Approximately two class periods.

Materials and Preparation

The items needed for this investigation are listed in Table 1.5. You may purchase additional beans, seeds, and nuts for this task or have students reuse the beans, seeds, and nuts that they dissected during Task 1. The other materials may be purchased through a science education supply company such as Ward's Science, Flinn Scientific, or Carolina Biological Supply. Print a set of Testing for Biomolecules Cards for each group; these can be downloaded from the book's Extras page at *www.nsta.org/mbi-biology*.

Table 1.5. Required Materials for Task 2

Item	Quantity
Safety goggles, nonlatex aprons, and vinyl or nitrile gloves	1 per student
Set of Testing for Biomolecules Cards	1 per group
Scalpel	1 per group
Mortar and pestle	1 per group
Well plate (12-well)	1 per group
Test tubes	1 per group
Test tube rack	1 per group
250-ml beaker	5 per group
Plastic pipettes	5 per group
1,000-ml Pyrex beaker (for hot water bath)	2 per class
Hot plate (for hot water bath)	2 per class
Test tube holder (for hot water bath)	2 per class
Dropper bottle of Benedict's solution	1 per group
Dropper bottle of Lugol's iodine solution	1 per group
Dropper bottle of biuret solution	1 per group
Brown paper cut into 5 × 5 cm squares	5 per group
Lima bean seeds	5 per group
Corn seeds	5 per group
Pine nuts (pine seeds)	5 per group
Sunflower seeds	5 per group
Pea seeds	5 per group

Continued

Table 1.5. Required Materials for Task 2 (*continued*)

Item	Quantity
Pinto bean seeds	5 per group
Kidney bean seeds	5 per group
Student Handout	1 per student

Place the materials in a central location in the classroom so students can get them during the task. You can set up two hot water baths in the classroom for students to use during the task. To make a hot water bath, simply add about 300 ml of water to a 100-ml Pyrex beaker and place on a hot plate. Set the hot plate to low so it will keep the water at about 158°F (70°C).

Safety Precautions

Remind students to follow all normal safety rules. In addition, tell students to take the following safety precautions:

- Wear sanitized indirectly vented chemical-splash goggles, nonlatex aprons, and vinyl or nitrile gloves during setup, hands-on activity, and cleanup.
- Be careful with scalpels. They are sharp and can cut or puncture skin.
- Use caution handling glass labware, which can shatter if dropped and cut skin or eyes.
- Never taste or eat any food items used in a lab activity.
- Use caution when working with hot plates, which can cause skin burns or electric shock.
- Use only GFI-protected circuits when using electrical equipment, and keep away from water sources to prevent shock.
- Use caution when working near hot water, which can splash and burn skin or eyes.
- Clean up any spills on the floor immediately to avoid a slip or fall hazard.
- Wash hands with soap and water when the activity is completed.

Procedure

This lesson plan is only a suggestion. It is included here to illustrate how you can facilitate student thinking during this task. We encourage you to modify this lesson plan by asking different questions, using different examples, and providing different scaffolds as appropriate to better meet the needs of students in your class.

Introduction of the Task (10 minutes)

1. Have students sit in groups of three or four.
2. Give each student one copy of the Student Handout for Task 2.

3. Read the Introduction and Initial Ideas sections of the handout aloud, having students follow along. It is a good idea to show a picture of General Sherman and seeds from a giant sequoia at this point in the lesson so students have the anchoring phenomenon of the unit in mind as they start this task.

4. Ask students to list the different types of biomolecules they think will be found in the endosperm of the seeds. Remind them to explain why they included each biomolecule on their list.

5. Give students an opportunity to share their ideas with the other people in their groups.

6. Ask one student from each group to share their ideas with the entire class.

Making Observations and Taking Measurements (60 minutes)

1. Read the section of the handout called Your Task aloud to students.

2. Show them the available materials and seeds they will use to complete this task. Ideally, students will use the endosperms they removed from the seeds during Task 1.

3. Go over the directions on the Testing for Biomolecules Cards. These cards include directions on how to use the different indicators to test for the presence or absence of each biomolecule in a sample. Don't tell students which types of seeds to test or how many different ones. Let them have a choice.

4. Go over the safety precautions for this task.

5. Give students about 40 minutes to use the indicators to test the endosperms of different seeds (see Box 1.1 for the testing procedure students should follow). Remind them to record what they observe or measure on their handouts.

6. As students work, move from group to group and check in with them. It's important to ask them questions that will help them connect what they are doing to the goal of the task and the anchoring phenomenon. (See the Back Pocket Questions for Task 2 on page 50 for some suggestions.)

7. Ask students to clean up their workstations and return the materials.

Box 1.1. Testing Procedure

To prepare the endosperm for testing:

1. Place three or four endosperms from one type of seed in the mortar. Only add the endosperms, not the seed coats or embryos.

2. Add 5 to 10 ml of water to the mortar.

3. Use the pestle to grind up the pieces of endosperm and mix them with a water. The mixture should have the consistency of a milkshake and should not contain any large chunks.

4. Transfer the mixture to a beaker.

To test for lipids in a sample:

1. Gather the sample, a pipette, and a piece of brown paper.

2. Use the pipette to put 1 drop of the sample on the brown paper.

3. Allow the drop to dry on the paper for at least 15 minutes.

To test for glucose (simple carbohydrates) in a sample:

1. Gather the sample, a test tube, a pipette, and a dropper bottle of Benedict's solution.

2. Use the pipette to add 10 drops of the sample to the test tube. (Do not use the same pipette for different samples.)

3. Add 20 drops of Benedict's solution to the same test tube.

4. Put the test tube in a hot water bath for 5 to 10 minutes.

To test for starch (complex carbohydrates) in a sample:

1. Gather the sample, a well plate, a pipette, and a dropper bottle of Lugol's iodine solution.

2. Use the pipette to add 5 drops of the sample to one of the wells in the well plate. (Do not use the same pipette for different samples.)

3. Add 5 drops of Lugol's iodine solution to the same well.

To test for protein in a sample:

1. Gather the sample, a well plate, a pipette, and a dropper bottle of biuret solution.

2. Use the pipette to add 10 drops of the sample to one of the wells in the well plate. (Do not use the same pipette for different samples.)

3. Add 20 drops of biuret solution to the same well.

Putting Ideas on the Table (10 minutes)

1. Give an interactive lecture to introduce some ideas that students might find useful for making sense of what they did. We recommend, as a minimum, introducing the following ideas:

    -   The molecular structures of sugars, starch, lipids, and proteins.
    -   The number of different bonds in these different molecules and how these bonds store chemical energy that is released when they are broken.

2. Encourage students to keep a record of the ideas they find important in the section called Some Useful Ideas From My Teacher on the Student Handout so they can refer to them later in the unit. Remember, your goal here is to put some ideas on the table to help your students make sense of what they are seeing and doing, not to tell them what they should have learned.

Adding Information to the Summary Table (10 minutes)

1. Give students 5 minutes to decide what to add to the Summary Table at the end of the Student Handout.

2. Have one student from each group share what the group figured out, how they know (their evidence for what they figured out), and how this information will help them explain the anchoring phenomenon.

3. Once each group has shared, ask the entire class to decide what should be added to each column of the class Summary Table for Task 2. Help students reach consensus about what to add to the Summary Table. Only add an idea to the Summary Table if everyone in the class agrees with that idea.

Back Pocket Questions

As students work in groups, it is important to engage with each group to help press and extend students' thinking around the ideas at play in this task. Following are some example questions you might ask:

1. Helping students get started: What biomolecules are present in that? How do you know?

2. Pressing further: Why might that biomolecule be found in the endosperm? What might be the function of the endosperm, since you know it contains that biomolecule?

3. Following up: What makes you think that? Can you say more?

Filling Out the Summary Table

Table 1.6 includes examples of the responses students may come up with when they fill out the Summary Table. This is provided here only as a behind-the-scenes roadmap and is not meant to be shared with students.

Table 1.6. Example Summary Table for Task 2

What we learned	How it helps us explain the phenomenon
The endosperm is made of starch, sugars, and fats. Starch, sugars, and fats are molecules that store chemical energy.	The endosperm could be an energy source for the embryo as it germinates because it is made up of starch and sugars.

Hints for Implementing This Task

- Many of the reagents used in this task can stain skin or clothes. You may wish to have students wear vinyl or nitrile gloves and nonlatex lab aprons.
- You may wish to show videos of the tests performed in this lab, which you can find on YouTube.

Possible Extensions

- In more advanced courses, students could also include a test for nucleic acids by performing a Dische diphenylamine test. This can be purchased through most science education supply companies.
- The test for lipids can be used to design an investigation into the lipid content of milk types (e.g., skim, 2%, whole).

Student Handout

TASK 2. WHAT IS THE FUNCTION OF THE ENDOSPERM?

Introduction

General Sherman is the largest tree in the world. A tree, like all plants, begins as a seed. You have already figured out what is inside a seed and what happens to a seed after it germinates. You know, for example, that the embryo inside a tree seed increases in size once it germinates and becomes a seedling, and the endosperm gets smaller over time. You now need to figure out the function of the endosperm in order to explain how General Sherman grew from an embryo to a seedling over time. One way to figure out what the endosperm does and why it gets smaller after a seed germinates is to determine what types of molecules make up the endosperm. This is important because the function of all living thing depends on the properties of its parts, so if you know the properties of the endosperm, you will be able to make inferences about what it might do.

Plants, like all living things, are made up of large, complex molecules called biological macromolecules or biomolecules. There are four main types of biomolecules: carbohydrates, lipids, proteins, and nucleic acids.

Carbohydrates are made up of carbon, hydrogen, and oxygen atoms. Sugars and starches are examples of molecules that are classified as carbohydrates. Living things use carbohydrates as a source of chemical energy. The chemical energy in carbohydrates is stored in the bonds that hold the atoms that make up the molecule together. Smaller molecules, such as sugars like glucose and sucrose, contain less chemical energy than larger molecules, such as starches, because smaller molecules are made up of fewer atoms and thus have fewer bonds than larger molecules. Plants also make cell structures out of carbohydrates.

Lipids are made up carbon, hydrogen, and oxygen atoms like carbohydrates, but they have a different molecular structure. Examples of lipids are fats and oils, waxes, and steroids. Living things use lipids to store chemical energy and also create cell structures out of lipids.

Proteins are large, complex molecules that carry out most functions of the cells and tissues found in plants. They are responsible for helping other molecules get in and out of cells, speeding up chemical reactions, producing other molecules, and breaking down waste. Proteins are made up of hundreds or thousands of smaller units called amino acids. There are 20 different types of amino acids, which are made up of carbon, hydrogen, oxygen, nitrogen, and sulfur atoms. The sequence of amino acids in a protein determines the function of the protein.

Nucleic acids are made up of carbon, hydrogen, oxygen, nitrogen, and phosphorus atoms. Plants, like animals, use nucleic acids to store or transfer information about how and when to make the many proteins needed to build and maintain functioning cells, tissues, and organisms. They also use nucleic acids to create new proteins, move amino acids around inside cells so they can be used to make new proteins, and regulate the process of creating proteins. Some examples of different types of biomolecules are illustrated in Figure 1.11.

Figure 1.11. Examples of Biomolecules

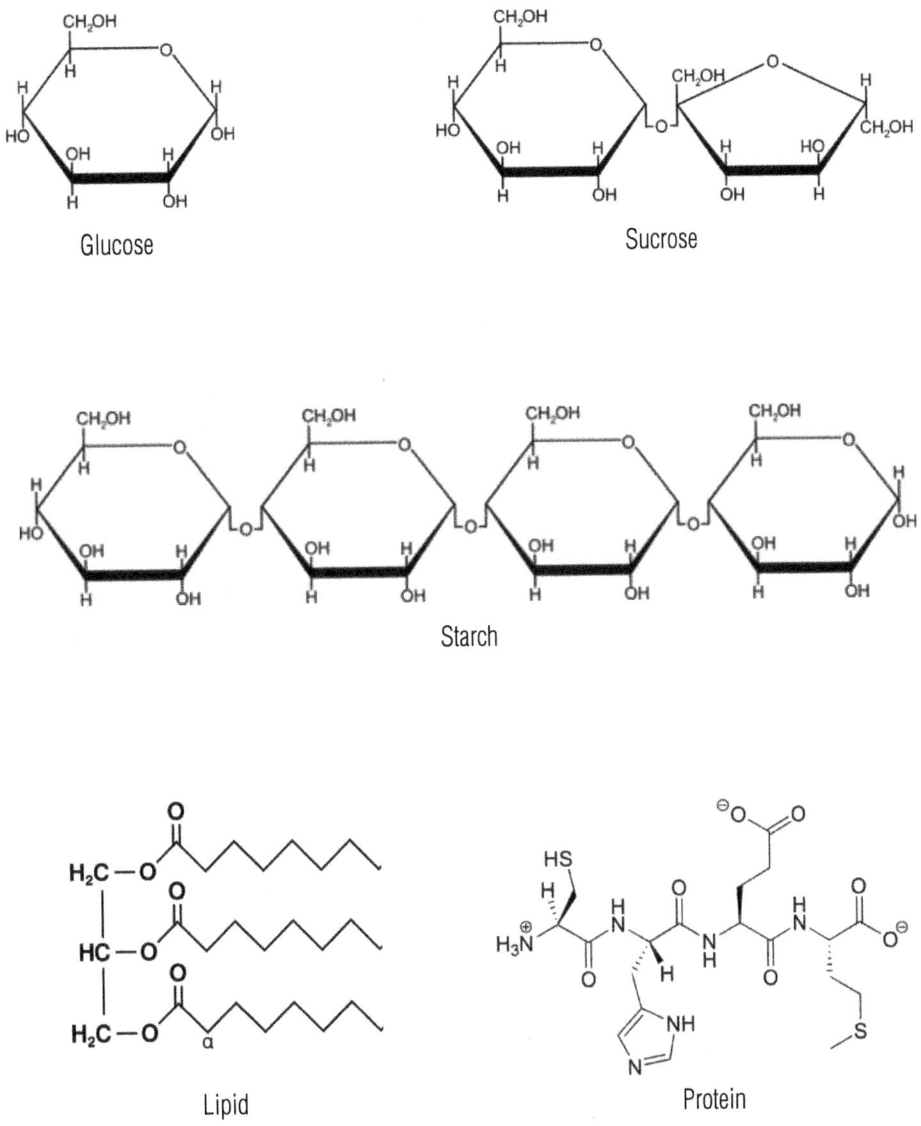

Glucose

Sucrose

Starch

Lipid

Protein

Initial Ideas

Before beginning this task, take a few minutes to think about what types of biomolecules might make up the endosperm. List the different types of biomolecules you **think** will be found in the endosperm of seeds in the space below. Be sure to explain why you included each biomolecule on your list.

Your Task

Your task is to use what you know about the relationship between *structure and function* in living things to *plan and carry out an investigation* to determine which biomolecules are found in the endosperm of different kinds of seeds. To accomplish this task, you will use chemical indicators to test for the presence or absence of carbohydrates (starch and glucose), lipids, and proteins. (Unfortunately, you will not have a way to test for nucleic acids.) Your teacher will show you how to use these chemical indicators before you begin. Be sure to keep track of anything you observe or measure as you work in the Observations and Measurements section of your handout.

Available Materials

You and your group may use any of the following materials during this task:

- Safety goggles, nonlatex aprons, and vinyl or nitrile gloves (required)
- Testing for Biomolecules Cards
- Scalpel
- Mortar and pestle
- Well plate (12-well)

- 6 test tubes
- Test tube rack
- Beaker
- Plastic pipettes
- Hot water bath
- Benedict's solution
- Lugol's iodine solution
- Biuret solution

- Brown paper
- Lima bean seeds
- Corn seeds
- Pine nuts (pine seeds)
- Sunflower seeds
- Pea seeds
- Pinto bean seeds
- Kidney bean seeds

Safety Precautions

Follow all normal lab safety rules. In addition, be sure to take the following safety precautions:

- Wear sanitized indirectly vented chemical-splash goggles, nonlatex aprons, and vinyl gloves during setup, hands-on activity, and cleanup.
- Be careful with scalpels. They are sharp and can cut or puncture skin.
- Use caution handling glass labware, which can shatter if dropped and cut skin or eyes.
- Never taste or eat any food items used in a lab activity.
- Use caution when working with hot plates, which can cause skin burns or electric shock.
- Use only GFI-protected circuits when using electrical equipment, and keep away from water sources to prevent shock.
- Use caution when working near hot water, which can splash and burn skin or eyes.
- Clean up any spills on the floor immediately to avoid a slip or fall hazard.
- Wash hands with soap and water when the activity is completed.

Observations and Measurements

Use the space below to keep a record of what you observe or measure as you work with the seeds.

Some Useful Ideas From My Teacher

You can keep track of useful ideas from your teacher in the space below.

What We Figured Out

Now that you have completed this task, take a few minutes to fill out the Summary Table on the following page with the other students in your group. This table will help you keep track of what you figured out during the task. You will then have an opportunity to share your ideas as we fill out our class Summary Table.

Summary Table

What we learned	How we know

How it helps us explain the phenomenon

Teacher Notes

Task 3. How Do Trees Use Sugar?

Purpose

This task gives students an opportunity to learn more about how plants get the energy they need to grow and carry out other essential functions. To accomplish this task, students test how carbon dioxide and oxygen levels change in a container that holds germinating seeds. This information helps students understand why the embryo in a seed gets bigger over time after germination but the endosperm gets smaller.

Important Life Science Content

One characteristic of all living things is that they must take in matter from the environment in order to survive. This is because all living tissues, which are composed of cells, need energy to function. Plants and animals get the energy they need to function from a chemical process that happens inside their cells called cellular respiration.

During cellular respiration, the bonds that hold sugar molecules and the bonds that hold oxygen molecules together are broken, and the atoms in these molecules are rearranged to create new molecules that transfer energy through a body and provide the energy needed to drive all the various functions of a cell. These molecules are called adenosine triphosphate (ATP). This chemical reaction also creates molecules of water and carbon dioxide as byproducts of the process. The oxygen needed for this chemical process comes from the environment. Animals get the sugar they need from the environment by eating plants or other animals. Plants get the sugar they need for cellular respiration through the process of photosynthesis (which is the focus of Tasks 4 and 5). The water and carbon dioxide are released back into the environment.

We can determine whether the process of cellular respiration is happening inside a living thing by tracking the amounts of oxygen and carbon dioxide that move into and out of the body of that living thing. A person, for example, breathes in air that is about 21% oxygen and less than 1% carbon dioxide and breathes out air that is about 15% oxygen and 5% carbon dioxide. There is less oxygen in the air that a person exhales because they absorbed some of the oxygen that was in the air they inhaled while that air was in their lungs. This oxygen is then transported from their lungs to individual cells throughout their body and used for cellular respiration. There is more carbon dioxide in the air that a person exhales because carbon dioxide moves from their cells to their lungs and then into the air. The air is then released into the environment as a waste product when the person exhales.

Scientific Ideas That Are Important to Think About During This Task

- Cellular respiration is a chemical process in which the bonds of sugar molecules and oxygen molecules are broken and new compounds are formed that can transport energy to cells and tissues throughout the plant.

- Germinating seeds take in oxygen and give off carbon dioxide just like animals do.

Timeline

Approximately two class periods.

Materials and Preparation

The items needed for this investigation are listed in Table 1.7. The pea seeds and plastic beads may be purchased from a big-box retail store such as Walmart or Target or through an online retailer such as Amazon. The sensors may be purchased from a company such as Vernier, PASCO, or PocketLab. The other materials will need to be purchased through a science education supply company such as Ward's Science, Flinn Scientific, or Carolina Biological Supply.

Table 1.7. Required Materials for Task 3

Item	Quantity
Safety goggles, nonlatex apron, and vinyl or nitrile gloves	1 per student
Oxygen sensor	1 per group
Carbon dioxide sensor	1 per group
Interface for the sensors	1 per group
250-ml Erlenmeyer flask	1 per group
Pea seeds	40 per group
500 ml of water	1 per class
Plastic beads (about the same size as a pea)	20 per group
Electronic scale	1 per group
1,000-ml beaker	3 per class
Student Handout	1 per student

To prepare enough materials for 10 groups of students to use, place 200 peas in one 1,000-ml beaker, another 200 peas in the second beaker, and 200 plastic beads in the third beaker. Add about 500 ml of water to one of the beakers of peas and allow to soak for at least 24 hours before this task begins. Right before class, place all the materials in a central location in the classroom so students can get them during the task.

Safety Precautions

Remind students to follow all normal safety rules. In addition, tell students to take the following safety precautions:

- Wear sanitized indirectly vented chemical-splash goggles, nonlatex aprons, and vinyl gloves during setup, hands-on activity, and cleanup.
- Use caution handling glass labware, which can shatter if dropped and cut skin or eyes.
- Clean up any spills on the floor immediately to avoid a slip or fall hazard.
- Wash hands with soap and water when the activity is completed.

Procedure

This lesson plan is only a suggestion. It is included here to illustrate how you can facilitate student thinking during this task. We encourage you to modify this lesson plan by asking different questions, using different examples, and providing different scaffolds as appropriate to better meet the needs of students in your class.

Introduction of the Task (10 minutes)

1. Have students sit in groups of three or four.
2. Give each student one copy of the Student Handout for Task 3.
3. Read the Introduction and Initial Ideas sections of the handout aloud, having students follow along. It is a good idea to show a picture of General Sherman and seeds from a giant sequoia at this point in the lesson so students have the anchoring phenomenon of the unit in mind as they start this task.
4. Ask students to draw and label on the handout how they think molecules of carbon dioxide and molecules of oxygen would move into or out of a seed before and after it germinates. Remind them to include labels that describe what is happening in their pictures.
5. Give students an opportunity to share their ideas with others in their group.
6. Ask one student from each group to share their ideas with the entire class.

Planning and Carrying Out the Investigation (60 minutes)

1. Read the section of the handout called Your Task aloud to students.
2. Show students the available materials and ask them how they might use some of the equipment. Don't be afraid to offer some suggestions, but do not require that every group do everything the same way. Give them a choice.
3. Go over the safety precautions for this task.
4. Give students about 15 minutes to plan out their investigation. As students work, move from group to group and check in with them. It's important to ask them questions that will help them connect what they are doing to the goal of the task and the anchoring phenomenon. (See the Back Pocket Questions for Task 3 on page 62 for some suggestions.)
5. Once the students have finished their investigation proposals, look them over to make sure they will be productive. If so, sign off on the proposal and let the students carry out their plan. If a proposal will not be productive, offer the students some advice about how

to address some of the weaknesses you see in their plan. Do not require everyone in the group to do the same thing. Give them a choice.

6. Give the students 40 minutes to collect the data they need.

7. As students work, again move from group to group and check in with them, asking questions such as those in the Back Pocket Questions on page 62.

8. Ask students to clean up their workstations and return the materials.

9. Once students have cleaned up their workstations, tell them to spend some time analyzing the data they collected before moving on to the next part of the lesson. Encourage students to think about how they make comparisons between conditions and what calculations they will need to do to identify a cause-and-effect relationship.

Putting Ideas on the Table (10 minutes)

1. Give an interactive lecture to introduce some ideas that students might find useful for making sense of what they did. We recommend, as a minimum, introducing the following ideas:

   - The number of different bonds in sugar molecules and how these bonds store chemical energy that is released when they are broken.

   - The fact that starch molecules consist of chains of sugar molecules and how starch molecules can be broken down and used as a source of sugar.

   - The fact that cellular respiration is a chemical process in which the bonds of sugar molecules and oxygen molecules are broken and new compounds are formed that can transport energy to cells and tissues throughout the plant.

   - The inputs (sugar and oxygen) and outputs (ATP, carbon dioxide, and water) of cellular respiration and where this process happens in cells (but *not* the specific details of the process, such as glycolysis, the Krebs cycle, and the electron transport chain or the exact numbers of ATP molecules that are created through this process).

2. Encourage students to keep a record of the ideas they find important in the section called Some Useful Ideas From My Teacher on the Student Handout so they can refer to them later in the unit. Remember, your goal here is to put some ideas on the table to help your students make sense of what they are seeing and doing, not to tell them what they should have learned.

Adding Information to the Summary Table (10 minutes)

1. Give students 5 minutes to decide what to add to the Summary Table at the end of the Student Handout.

2. Have one student from each group share what the group figured out, how they know (their evidence for what they figured out), and how this information will help them explain the anchoring phenomenon.

3. Once each group has shared, ask the entire class to decide what should be added to each column of the class Summary Table for Task 3. Help students reach consensus about what to add to the Summary Table. Only add an idea to the Summary Table if everyone in the class agrees with that idea.

Back Pocket Questions

As students work in groups, it is important to engage with each group to help press and extend students' thinking around the ideas at play in this task. Following are some example questions you might ask:

1. Helping students get started: Why is it important to track changes in oxygen or carbon dioxide in the air? Why would it be useful to use the Erlenmeyer flask to create a closed system to collect the data you need?

2. Pressing further: How would you know cellular respiration is taking place inside a seed? What potential cause-and-effect relationship are you looking for? What information do you need and what types of conditions do you need to set up to identify a cause-and-effect relationship?

3. Following up: What makes you think that? Can you say more?

Filling Out the Summary Table

Table 1.8 includes examples of the responses students may come up with when they fill out the Summary Table. This is provided here only as a behind-the-scenes roadmap and is not meant to be shared with students.

Table 1.8. Example Summary Table for Task 3

What we learned	How it helps us explain the phenomenon
Cellular respiration is a chemical process in which the bonds of sugar molecules and oxygen molecules are broken and new compounds are formed that can transport energy to cells and tissues throughout the plant.	The embryo will use the starch and sugar in the endosperm for cellular respiration. Once the endosperm is used up, there is not a source of sugar for cellular respiration.

Hints for Implementing This Task

- This is a very open lab. Students may need additional support with identifying possible variables to test (e.g., temperature, germinating versus nongerminating seeds).

- It is important that students include controls in their design. Asking groups to describe their research questions and procedures may help them identify problems in their own designs.

Possible Extensions

- After the first investigation, students could design further investigations to test for the effects of additional variables.
- There are important connections to agriculture and gardening that students can learn about, especially the importance of temperature to germination.

Student Handout

TASK 3. HOW DO TREES USE SUGAR?

Introduction

One characteristic of living things is that they must take in matter from the environment to survive. This is because all living tissues, which are composed of cells, need energy to function. Animals get the energy they need to function from a chemical process that occurs inside their cells called *cellular respiration*. During cellular respiration, the bonds that hold sugar molecules and the bonds that hold oxygen molecules together are broken, and the atoms in these molecules are rearranged to create new molecules that can provide the energy needed to drive all the various functions of a cell. This chemical reaction also creates molecules of water and carbon dioxide as part of the process.

The sugar and oxygen needed for this chemical process come from the environment. Water and carbon dioxide, which are waste products, are released back into the environment. You can therefore determine whether the process of cellular respiration is happening inside a living thing by tracking the amounts of oxygen and carbon dioxide that move into and out of it. A person, for example, breathes in air that is about 21% oxygen and less than 1% carbon dioxide and breathes out air that is about 15% oxygen and 5% carbon dioxide. There is less oxygen in the air that a person exhales because they absorbed some of the oxygen that was in the air they inhaled while that air was in their lungs. This oxygen is then transported from their lungs to individual cells throughout their body and used for cellular respiration. There is more carbon dioxide in the air that a person exhales because carbon dioxide moves from their cells to their lungs and then into the air. The air is then released into the environment as a waste product when the person exhales.

All animals use this process to turn the chemical energy found in the food they eat into a form that they can use to move, grow, feed, and reproduce. It is a unifying characteristic of all animals. But what about other types of living things, like plants? Do plants also break the bonds that hold the atoms of sugar molecules and oxygen molecules together and then create new molecules that can transport energy throughout the plant?

You already generated some evidence during the last two tasks that seems to support the idea that plants use sugar as a source of energy like animals do. For example, you know that the seeds of trees, such as General Sherman, contain an embryo and a structure called an endosperm. You also know that after a seed germinates, the endosperm, which is made up of starch, sugar, and lipids, will decrease in size over time as the embryo increases in size. In this task, you need to figure out whether seeds take in oxygen and give off carbon dioxide after they germinate to determine whether cellular respiration happens inside the cells of plants like it does inside the cells of animals.

Initial Ideas

Before beginning this task, take a few minutes to think about how matter, such as molecules of oxygen and carbon dioxide, moves into and out of a germinating seed if cellular respiration is happening inside its cells. In the boxes below, draw and label how you think molecules of carbon dioxide and oxygen would move into or out of a seed before and after it germinates. Be sure to include labels that describe what is happening.

Movement of Carbon Dioxide and Oxygen

BEFORE germinating	AFTER germinating

Your Task

Your task is to use what you know about *tracking the movement of matter into, within, and out of* living things and *cause-and-effect relationships* to *plan and carry out an investigation* to determine whether plants, like animals, get the energy they need to function from a chemical process called cellular respiration.

Available Materials

You and your group may use any of the following materials during this task:

- Safety goggles, nonlatex apron, and vinyl or nitrile gloves (required)
- Oxygen sensor
- Carbon dioxide sensor
- Interface for the sensors
- 250-ml Erlenmeyer flask
- Germinating pea seeds
- Dry pea seeds
- Plastic beads
- Electronic scale

Safety Precautions

Follow all normal lab safety rules. In addition, be sure to take the following safety precautions:

- Wear sanitized indirectly vented chemical-splash goggles, nonlatex apron, and vinyl gloves during setup, hands-on activity, and cleanup.
- Use caution handling glass labware, which can shatter if dropped and cut skin or eyes.
- Clean up any spills on the floor immediately to avoid a slip or fall hazard.
- Wash hands with soap and water when the activity is completed.

Plan Your Investigation

First, develop a plan by filling out an investigation proposal. Then have your teacher look it over and approve it before beginning your investigation.

Investigation Proposal

The *question* we are trying to answer:

We will collect the following *data*:

This is a drawing of how we will *set up the equipment*:

This is what we will do to *collect the data* we need:

These are the *safety precautions* we will follow:

We will *analyze* the data we collect by:

I approve of this investigation.

_____   _____
Teacher's Signature   Date

National Science Teaching Association

Observations and Measurements

Use the space below to keep a record of the data you collect during your investigation.

Analyze Your Data

Create a graph in the space below to help identify any cause-and-effect relationships. You may need to make some calculations before you can make a graph.

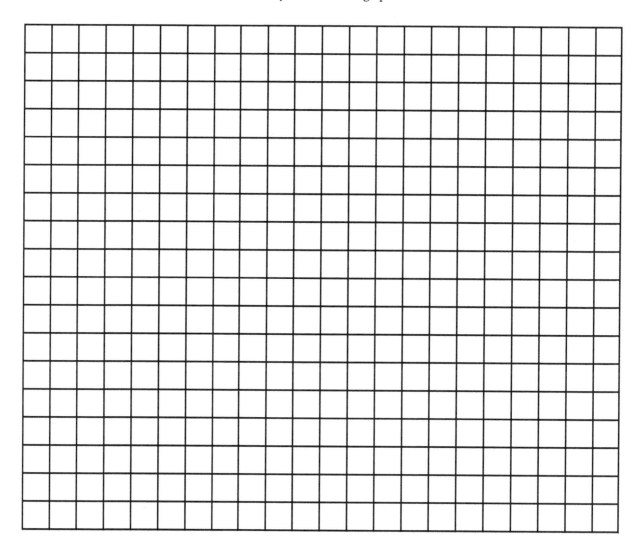

Some Useful Ideas From My Teacher

You can keep track of useful ideas from your teacher in the space below.

What We Figured Out

Now that you have completed this task, take a few minutes to fill out the Summary Table on the following page with the other students in your group. This table will help you keep track of what you figured out during the task. You will then have an opportunity to share your ideas as we fill out our class Summary Table.

Summary Table

What we learned	How we know

How it helps us explain the phenomenon

Teacher Notes

Task 4. How Does the Endosperm Produce Sugar?

Purpose

This task gives students an opportunity to learn more about how plants create the sugar they need for cellular respiration. To accomplish this task, students analyze and interpret data collected by a biologist named James Smith during a series of experiments that he carried out in the late 1930s–40s. Performing this task helps students figure out which molecules plants take in and convert into the sugar they need for cellular respiration.

Important Life Science Content

The process of photosynthesis converts light energy to stored chemical energy by converting carbon dioxide plus water into sugars plus released oxygen. In this task, students must figure out the inputs and outputs of photosynthesis by analyzing data that were generated based on the results of a series of experiments conducted by James Smith in the late 1930s–40s on sunflowers and reported by Smith (1940, 1943, 1944) and Smith and Cowie (1941). Although the data that the students will analyze during the task are not the actual measurements Smith collected, they are intended to be consistent with his findings. We have also added some data to promote and support student sensemaking during this task.

Students can determine the inputs and outputs of photosynthesis by tracking the movement of matter into, out of, or within a plant under different conditions. Students will be able to examine the production of carbohydrates in sunflower leaves under six different conditions. Each condition represents a different test:

1. High moisture, no light, and a normal carbon dioxide concentration in the air
2. High moisture, no light, and a high carbon dioxide concentration in the air
3. Low moisture, no light, and a normal carbon dioxide concentration in the air
4. High moisture, light, and a normal carbon dioxide concentration in the air
5. High moisture, light, and a high carbon dioxide concentration in the air
6. Low moisture, light, and a high carbon dioxide concentration in the air

The results of tests 1–3 show that the creation of carbohydrates requires light. These tests also show that the sunflower leaves are alive, because the change in the oxygen level (measured in parts per thousand [ppt]) of the air inside the container holding the leaves decreases during this test, while the carbon dioxide levels (also measured in ppt) increase. These changes in oxygen and carbon dioxide levels indicate that cellular respiration is occurring inside the cells of the leaves, even though photosynthesis is not. The results of tests 4–6 show that in the presence of light, sunflower leaves produce carbohydrates, and they use carbon dioxide and give off oxygen as part of this process. Students can conclude that the plants use carbon dioxide from the air for this process because the carbon dioxide level of the air inside the container holding the leaves decreases and the nitrogen level does not change. The increase in the oxygen level in the containers also suggests that oxygen is given off as a byproduct of the process. Finally, the results of tests 5 and 6 indicate that water is needed for this reaction, because more carbohydrates are produced in the high-moisture condition than in the low-moisture condition. Presumably, the leaves in the low-moisture condition used up all the available water, so photosynthesis ceased.

Scientific Ideas That Are Important to Think About During This Task

- The process of photosynthesis converts light energy to stored chemical energy by converting carbon dioxide plus water into sugars plus released oxygen.

Timeline

Approximately two class periods.

Materials and Preparation

The items needed for this investigation are listed in Table 1.9.

Table 1.9. Required Materials for Task 4

Item	Quantity
Calculator	1 per group
2 × 3 foot dry-erase whiteboard	1 per group
Dry-erase pens	1 set per group
Student Handout	1 per student

Safety Precautions

This task does not require any specific safety precautions.

Procedure

This lesson plan is only a suggestion. It is included here to illustrate how you can facilitate student thinking during this task. We encourage you to modify this lesson plan by asking different questions, using different examples, and providing different scaffolds as appropriate to better meet the needs of students in your class.

Introduction of the Task (10 minutes)

1. Have students sit in groups of three or four.
2. Give each student one copy of the Student Handout for Task 4.
3. Read the Introduction and Initial Ideas sections of the handout aloud, having students follow along. It is a good idea to show a picture of General Sherman and seeds from a

giant sequoia at this point in the lesson so students have the anchoring phenomenon of the unit in mind as they start this task.

4. Ask students to draw and label on the handout how they think seedlings produce the sugar they need to keep growing once the endosperm is used up. Remind them to include labels that describe what is happening in their pictures.

5. Give students an opportunity to share their ideas with the others in their group.

6. Ask one student from each group to share their ideas with the entire class.

Analyzing and Interpreting Data and Then Arguing From Evidence (60 minutes)

1. Read the section of the handout called Your Task aloud to students.

2. Go over the tests conducted by James Smith and show students the results of the tests. Be sure to ask questions about how they might make sense of the data. Don't be afraid to offer some suggestions, but do not require that every group do everything the same way. Give them a choice.

3. Give students about 20 minutes to analyze the data. As students work, move from group to group and check in with them. It's important to ask them questions that will help them connect what they are doing to the goal of the task and the anchoring phenomenon. (See the Back Pocket Questions on pages 78–79 for some suggestions.)

4. Go over the section called Your Argument with students. Tell them that they need to create an argument on a whiteboard to share with others what they figured out and their evidence in support of what they figured out. They also need to write down things they are unsure about on the board so their classmates can provide them with feedback.

5. Give students 15 minutes to create an argument on a whiteboard.

6. As students work, again move from group to group and check in with them, asking questions such as those in the Back Pocket Questions on pages 78–79.

7. Give students about 10 minutes to share, evaluate, and revise their arguments. We recommend using a modified gallery walk format for this part of the task. In this format, one or two members of the group (the "presenters") stay at their workstation to share their group's ideas, while the other group members (the "travelers") go to different groups one at a time to listen to and critique the arguments developed by their classmates. We recommend that the travelers visit at least three different workstations during the argumentation session, and that the presenters keep a record of the critiques made by their classmates and any suggestions for improvement. The travelers should also be encouraged to keep a record of good ideas or potential ways to improve their own arguments as they visit different groups.

8. Give students about 10 minutes to meet with their original groups so they can discuss what they learned by interacting with individuals from other groups and can revise their initial arguments. This process can begin with the presenters sharing the critiques and suggestions for improvement that they heard during the argumentation session. The

students who visited other groups can then share their ideas for making the arguments better based on what they observed and discussed at other stations.

9. Have students clean up their workstations and then move on to the next part of the lesson.

Putting Ideas on the Table (10 minutes)

1. Give an interactive lecture to introduce some ideas that students might find useful for making sense of what they did. We recommend, as a minimum, introducing or reminding students of the following ideas:

   - The main way that solar energy is captured and stored on Earth is through the complex chemical process known as photosynthesis.
   - The total amount of energy and matter in closed systems is conserved.
   - Changes in the matter within a system can be identified by tracking the amount of matter flowing into or out of that system.

2. Encourage students to keep a record of the ideas they find important in the section called Some Useful Ideas From My Teacher on the Student Handout so they can refer to them later in the unit. Remember, your goal here is to put some ideas on the table to help your students make sense of what they are seeing and doing, not to tell them what they should have learned.

Adding Information to the Summary Table (10 minutes)

1. Give students 5 minutes to decide what to add to the Summary Table at the end of the Student Handout.

2. Have one student from each group share what the group figured out, how they know (their evidence for what they figured out), and how this information will help them explain the anchoring phenomenon.

3. Once each group has shared, ask the entire class to decide what should be added to each column of the class Summary Table for Task 4. Help students reach consensus about what to add to the Summary Table. Only add an idea to the Summary Table if everyone in the class agrees with that idea.

Back Pocket Questions

As students work in groups, it is important to engage with each group to help press and extend students' thinking around the ideas at play in this task. Following are some example questions you might ask:

1. Helping students get started: How would you know photosynthesis is taking place inside the leaves based on the data available in the table? How would you know if a plant is using a molecule for photosynthesis or giving off a molecule as a result of photosynthesis?

2. Pressing further: Why is it important to track changes in the nitrogen, oxygen, and carbon dioxide in the air? What types of calculations could you make to help identify any patterns in these data? What types of mathematical representations could you create to help identify any patterns in these data?

3. Following up: What makes you think that? Can you say more?

Filling Out the Summary Table

Table 1.10 includes examples of the responses students may come up with when they fill out the Summary Table. This is provided here only as a behind-the-scenes roadmap and is not meant to be shared with students.

Table 1.10. Example Summary Table for Task 4

What we learned	How it helps us explain the phenomenon
The main way that solar energy is captured and stored on Earth is through a chemical process known as photosynthesis. This process converts light energy to stored chemical energy by converting carbon dioxide plus water into sugars plus released oxygen.	General Sherman is able to convert light energy to stored chemical energy by converting carbon dioxide plus water into sugars plus released oxygen. These sugars are then used for cellular respiration. General Sherman is able to produce all the sugar it needs to keep growing.

Hints for Implementing This Task

- Some students may need assistance with interpreting the data table and producing the graph.

Possible Extensions

- The history of the discovery of photosynthesis dates to the 17th century and is quite interesting. A brief summary can be found at *https://photosynthesiseducation.com/discovery-of-photosynthesis*.

References

Smith, J. H. C. 1940. The absorption of carbon dioxide by unilluminated leaves. *Plant Physiology* 15 (2): 183–224.

Smith, J. H. C. 1943. Molecular equivalence of carbohydrates to carbon dioxide in photosynthesis. *Plant Physiology* 18 (2): 207–223.

Smith, J. H. C. 1944. Concurrency of carbohydrate formation and carbon dioxide absorption during photosynthesis in sunflower leaves. *Plant Physiology* 19 (3): 394–403.

Smith, J. H. C., and D. R. Cowie. 1941. Absorption and utilization of radioactive carbon dioxide by sunflower leaves. *Plant Physiology* 16 (2): 257–271.

Student Handout

TASK 4. HOW DOES THE ENDOSPERM PRODUCE SUGAR?

Introduction

All living things need energy in order to function. Plants, like animals, get the energy they need to function from *cellular respiration*, during which the atoms of sugar and oxygen molecules are rearranged to create new molecules that provide the energy needed for cell functions. Molecules of water and carbon dioxide are created as byproducts of the process.

Animals get the oxygen and sugar they need for cellular respiration from the environment. They can obtain the oxygen they need by absorbing it through their skin, lungs or gills, but they must eat other living things that contain carbohydrates to get the sugar they need for cellular respiration. But what about plants, which do not eat other living things? We know that the embryo in a seed can use the sugar in the endosperm for cellular respiration as it grows into a seedling. We also know that this source of sugar is quickly used up as the plant grows larger over time. So how do plants get the sugar they need once the endosperm is gone?

Biologists have long known that plants are able to produce sugar and store the excess sugar that they make in their tissues as starch. The carbohydrates that are stored in the stems, leaves, roots, and seeds of plants allow animals to obtain the sugar they need to survive by eating plants. It took a long time and a great deal of research for biologists to figure out the exact chemical process that plants use to produce their own carbohydrates. In this task, you will have an opportunity to examine some data collected by a biologist named James Smith during a series of experiments he carried out in the late 1930s–40s using the leaves of sunflowers. Your goal is to figure out which molecules a seedling uses to produce the carbohydrates they need for cellular respiration once they use up the sugar found in the endosperm.

Initial Ideas

Before beginning this task, take a few minutes to think about how a seedling might be able to produce the sugar it needs for cellular respiration. In the space below, draw and label how you think a seedling produces the sugar it needs for cellular respiration after the sugar in the endosperm of a seed is used up. Be sure to include labels that describe what is happening.

Your Task

Your task is to use what you know about *matter and energy flow in living things* and the *transfer of energy and matter into and out of systems* to *analyze and interpret data* published by James Smith between 1940 and 1944 to determine which molecules a plant uses to produce the sugar it needs for cellular respiration. Then craft an argument to share what you have figured out with your classmates and explain how you know what you figured out is valid or acceptable.

Tests Conducted by James Smith

James Smith cut squares from sunflower leaves to create "leaves" that were the same size and thickness. Then he soaked some of the leaf squares in water for 30 minutes so they were well hydrated and dried out the rest of them so there was very little water left inside these squares. He put the leaf squares inside a sealed container to create a closed system in which he could change specific environmental conditions, such as the amount of time in light and the amount of carbon dioxide in the air. After a set amount of time, he observed and measured changes in the leaf squares. He was interested in finding out how much glucose, sucrose, and starch a leaf would produce under the following conditions:

1. High moisture, no light, and a normal carbon dioxide concentration in the air
2. High moisture, no light, and a high carbon dioxide concentration in the air
3. Low moisture, no light, and a normal carbon dioxide concentration in the air
4. High moisture, light, and a normal carbon dioxide concentration in the air
5. High moisture, light, and a high carbon dioxide concentration in the air
6. Low moisture, light, and a high carbon dioxide concentration in the air

Results of the Tests

Table SH4.1 includes some data that James Smith collected during his tests on sunflower leaves. Assume that the conditions in the containers that are not listed were identical across all 30 tests.

Table SH4.1. Data From Tests on Sunflower Leaves

| Test | Conditions | | | | | Observations and measurements | | | | | | |
| | Time in light (min.) | Moisture level | Initial gas concentration (ppt) | | | Final gas concentration (ppt) | | | Amount of carbohydrate produced (mg) | | | |
			Nitrogen	Oxygen	Carbon dioxide	Nitrogen	Oxygen	Carbon dioxide	Glucose	Sucrose	Starch	Total
1a	0	High	780.7	209.4	0.4	780.8	201.2	2.7	0.1	0.1	0.0	0.2
1b	0	High	780.6	209.2	0.4	780.6	201.1	2.5	0.3	0.1	0.0	0.4
1c	0	High	780.7	209.3	0.4	780.7	201.2	2.7	0.2	0.0	0.0	0.2
1d	0	High	780.7	209.4	0.4	780.7	201.3	2.8	0.1	0.0	0.0	0.1
1e	0	High	780.7	209.4	0.3	780.7	201.3	2.6	0.1	0.1	0.0	0.1
2a	0	High	780.7	209.4	37.0	780.7	201.4	39.2	0.1	0.1	0.1	0.3
2b	0	High	780.6	209.4	37.1	780.6	201.5	39.3	0.2	0.1	0.1	0.4
2c	0	High	780.7	209.3	36.9	780.7	201.2	39.4	0.2	0.0	0.0	0.2
2d	0	High	780.7	209.4	37.0	780.7	201.2	39.5	0.1	0.1	0.0	0.2
2e	0	High	780.8	209.5	37.0	780.8	201.3	39.6	0.1	0.1	0.1	0.3
3a	0	Low	780.7	209.4	0.4	780.7	202.7	2.1	0.1	0.1	0.0	0.2
3b	0	Low	780.8	209.4	0.4	780.8	202.8	2.2	0.1	0.1	0.0	0.2
3c	0	Low	780.8	209.5	0.4	780.8	202.7	2.3	0.2	0.1	0.0	0.3
3d	0	Low	780.7	209.5	0.3	780.7	202.6	2.3	0.1	0.0	0.0	0.1
3e	0	Low	780.7	209.4	0.4	780.7	202.6	2.1	0.1	0.2	0.1	0.4
4a	60	High	780.7	209.4	0.4	780.7	213.6	0.0	0.2	2.3	1.5	4.0
4b	60	High	780.7	209.4	0.4	780.7	213.6	0.0	0.2	2.4	1.5	4.1
4c	60	High	780.7	209.3	0.4	780.7	213.7	0.0	0.3	2.6	1.3	4.2
4d	60	High	780.6	209.4	0.4	780.6	213.5	0.0	0.3	2.3	1.2	3.8
4e	60	High	780.7	209.5	0.4	780.7	213.6	0.0	0.3	2.2	1.9	4.4

Continued

Table SH4.1. Data From Tests on Sunflower Leaves (*continued*)

Test	Conditions						Observations and measurements								
	Time in light (min.)	Moisture level	Initial gas concentration (ppt)			Final gas concentration (ppt)			Amount of carbohydrate produced (mg)			
			Nitrogen	Oxygen	Carbon dioxide	Nitrogen	Oxygen	Carbon dioxide	Glucose	Sucrose	Starch	Total
5a	60	High	780.7	209.4	37.0	780.7	225.6	32.7	2.5	7.1	4.3	13.9
5b	60	High	780.7	209.5	37.1	780.7	225.6	32.8	2.6	7.0	4.3	13.9
5c	60	High	780.7	209.4	37.0	780.7	225.5	32.8	2.4	7.2	4.5	14.1
5d	60	High	780.6	209.4	36.9	780.6	225.5	32.6	2.5	7.0	4.2	13.7
5e	60	High	780.6	209.3	36.9	780.6	225.7	32.6	2.4	7.2	4.4	14.0
6a	60	Low	780.7	209.4	37.0	780.7	217.6	34.7	.9	4.2	2.0	7.1
6b	60	Low	780.7	209.2	37.1	780.7	217.5	34.8	.8	4.0	2.6	7.4
6c	60	Low	780.8	209.4	37.1	780.8	217.7	34.6	1.0	3.9	2.1	7.0
6d	60	Low	780.7	209.4	37.0	780.7	217.7	34.7	1.1	4.1	2.2	7.4
6e	60	Low	780.7	209.5	36.9	780.7	217.5	34.8	1.2	4.2	1.8	7.2

Source: The authors generated these data based on tables and graphs reported in Smith (1940, 1943, 1944) and Smith and Cowie (1941). The data are intended for educational use only.

Analyze and Interpret the Data

Use the space below to create some graphs to help make sense of the results of the experiments. You may need to make several calculations as part of your analysis.

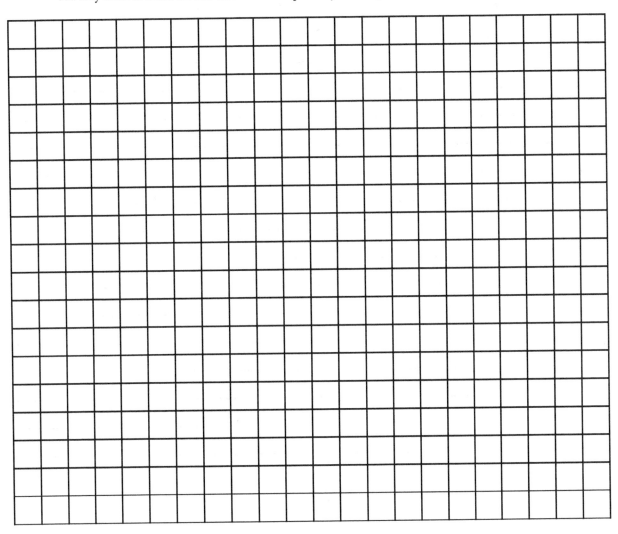

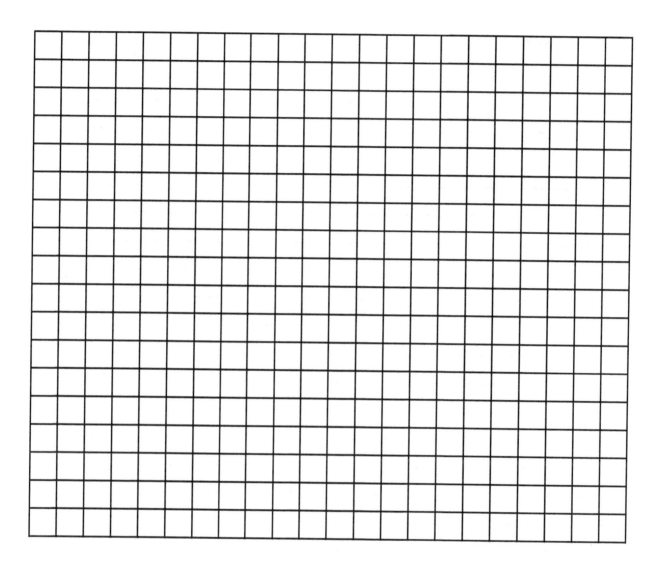

National Science Teaching Association

Your Argument

Develop an argument on a whiteboard. It should include these three components:

1. *What you figured out:* State a claim about how or why something happens.

2. *Your evidence:* Support your claim with an analysis of the available data and an interpretation of what the analysis means.

3. *Things you are unsure about:* List anything your group would like to learn more about in order to develop a more complete claim or to provide stronger evidence for your claim.

The Guiding Question:	
Our Claim:	
Our Evidence:	Our Justification of the Evidence:

Share your argument with your classmates. Be sure to ask them how to make your draft argument better. Keep track of their suggestions in the space below.

Ways to IMPROVE our argument …

Some Useful Ideas From My Teacher

You can keep track of useful ideas from your teacher in the space below.

What We Figured Out

Now that you have completed this task, take a few minutes to fill out the Summary Table below with the other students in your group. This table will help you keep track of what you figured out during the task. You will then have an opportunity to share your ideas as we fill out our class Summary Table.

Summary Table

What we learned	How we know

How it helps us explain the phenomenon

References

Smith, J. H. C. 1940. The absorption of carbon dioxide by unilluminated leaves. *Plant Physiology* 15 (2): 183–224.

Smith, J. H. C. 1943. Molecular equivalence of carbohydrates to carbon dioxide in photosynthesis. *Plant Physiology* 18 (2): 207–223.

Smith, J. H. C. 1944. Concurrency of carbohydrate formation and carbon dioxide absorption during photosynthesis in sunflower leaves. *Plant Physiology* 19 (3): 394–403.

Smith, J. H. C., and D. R. Cowie. 1941. Absorption and utilization of radioactive carbon dioxide by sunflower leaves. *Plant Physiology* 16 (2): 257–271.

Midunit Model Revision for Unit 1

Purpose

Now that your class is roughly halfway through the unit, it is important for students to go back to the initial models constructed on the first day and revise them based on what they have learned so far. Students may choose between two different strategies for model revisions. After negotiating what should be added, revised, or removed from their models, the groups may do either of the following:

- Redraw their models.
- Use sticky notes to keep track of their revisions for the final model revision near the end of the unit.

They should make the choice between these options based on the amount of time available for the model revision process. It is important, however, that students make decisions about revisions based on what fits with the evidence from the tasks or is consistent with the scientific ideas at play. We have provided a model revision PowerPoint template for your use, which can be downloaded from the book's Extras page at *www.nsta.org/mbi-biology*, and Figure 1.12 contains resources to help with modeling.

Figure 1.12. Resources to Help With Modeling

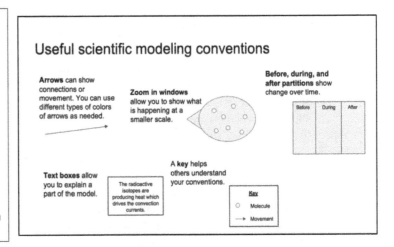

Although it does take time, we recommend a share-out session after the model revision. This allows for new ideas to be shared across the whole class, as well as discussion about what should and should not be included in the models. Have the members of each group stand up and describe how and why they changed their model. If time is limited, you could ask each group to focus on just one change. Provide time for questions from the class, and be sure to ask clarifying questions while comparing and contrasting the ideas presented across the groups.

Teacher Notes

Task 5. How Do Plants Grow?

Purpose

This task gives students an opportunity to learn more about how plants increase in size through the process of cell division. To accomplish this task, students examine the cells of an onion root tip and then watch several videos that show what happens inside animals as their cells divide. These ideas help students understand how large multicellular organisms, such as General Sherman, increase in size over time.

Important Life Science Content

Cells spend most of their lives growing, though bodies of multicellular organisms still have cells that are incredibly small. This is possible through cell division, known as mitosis, in which the body grows by producing more cells called daughter cells.

Cells must divide or reproduce for a number of reasons. They must divide if they get too big. As a cell gets larger, its surface area–to–volume ratio increases, creating more space within the cell for molecules to travel. This means substances must travel farther to reach where they are needed, which is incredibly inefficient. ATP is used during most active transport of molecules and ions within the cell; the larger the cell, the more ATP must be used. When a cell divides, it reduces its surface area–to–volume ratio, becoming more efficient.

Larger cells require more proteins and maintenance to function. Each protein needed within a cell is created by reading sections of DNA, creating messenger RNA (mRNA), and building proteins. However, DNA cannot be read fast enough to support larger cells that require large numbers of proteins. ATP is necessary in the assembly of proteins, and thus larger cells that require larger numbers of proteins are less efficient.

Cells also divide in order to make new cells. When part of an organism has become damaged, existing cells divide to fill the space to repair or heal the organism. Similarly, new cells must be created in order for a multicellular organism to grow, since new cells are smaller and more efficient. Plant and animal cells divide through the process of mitosis, in which two identical daughter cells are created from one original cell. All newly formed cells require DNA, so before a cell divides, a copy of DNA is made for each daughter cell. Additionally, all newly formed cells require organelles. The original cell splits up its organelles evenly between the two daughter cells. Daughter cells must then use ATP to build up organelles and grow so they can divide again. Mitosis is an active process; therefore, ATP is used continuously.

Scientific Ideas That Are Important to Think About During This Task

- In plants, like all multicellular organisms, individual cells grow and then divide via a process called mitosis, thereby allowing the organism to grow.

Timeline

Approximately one class period.

Materials and Preparation

The items needed for this investigation are listed in Table 1.11. The microscope and slide may be purchased from a science education supply company such as Ward's Science, Flinn Scientific, or Carolina Biological Supply. Place the materials in a central location in the classroom so students can get them during the task.

Table 1.11. Required Materials for Task 5

Item	Quantity
Prepared slide of an onion root	2 per group
Microscope	2 per group
Computer or tablet with internet access	1 per group
Student Handout	1 per student

Safety Precautions

Remind students to follow all normal safety rules. In addition, tell students to take the following safety precautions:

- Use only GFI-protected circuits when using electrical equipment, and keep away from water sources to prevent shock.
- Wash hands with soap and water when the activity is completed.

Procedure

This lesson plan is only a suggestion. It is included here to illustrate how you can facilitate student thinking during this task. We encourage you to modify this lesson plan by asking different questions, using different examples, and providing different scaffolds as appropriate to better meet the needs of students in your class.

Introduction of the Task (10 minutes)

1. Have students sit in groups of three or four.
2. Give each student one copy of the Student Handout for Task 5.
3. Read the Introduction and Initial Ideas sections of the handout aloud, having students follow along. It is a good idea to show a picture of General Sherman and seeds from a giant sequoia at this point in the lesson so that students have the anchoring phenomenon of the unit in mind as they start this task.
4. Ask students to draw on the handout the structures they think are inside a root and how these structures change as the roots grow on their own. Remind them to describe the functions of those structures in their pictures.

5. Give students an opportunity to share their pictures with the others in their group.

6. Ask one student from each group to share how their ideas about how a root grows were similar or different from the ideas of others in the group.

Making Observations and Taking Measurements (30 minutes)

1. Read the section of the handout called Your Task aloud to students.

2. Show them the available materials. Be sure show them how to use a microscope, but do not show them the onion root slides that they will use during the task.

3. Go over the safety precautions for this task.

4. Give students about 15 minutes to make observations about the cells found in the tip of an onion root. Remind them to record what they observe or measure on their handouts.

5. As students work, move from group to group and check in with them. It's important to ask them questions that will help them connect what they are doing to the goal of the task and the anchoring phenomenon. (See the Back Pocket Questions for Task 5 on page 95 for some suggestions.)

6. Ask students to clean up their workstations and return the materials.

7. Give students about 10 minutes to watch the three videos of animal cell division, which occurs inside an animal as it grows. Remind them to be sure to record what they observe on their handouts.

Putting Ideas on the Table (10 minutes)

1. Give an interactive lecture to introduce some ideas that students might find useful for making sense of what they did. We recommend, as a minimum, introducing the following ideas:

   - The process of cell division (emphasize the sequence of events, not the names of the stages).

   - The fact that each new cell that is created has the same DNA as the original cell so the whole plant has the information it needs to function and keep growing.

   - The fact that cells cannot grow in size beyond a certain limit. If a cell gets too large, materials such as gases and food molecules will not be able to cross the cell membrane fast enough to keep the cell alive. When it gets too large, the cell must divide into smaller cells to increase its surface area–to–volume ratio or it will cease to function.

2. Encourage students to keep a record of the ideas they find important in the section called Some Useful Ideas From My Teacher on the Student Handout so they can refer to them later in the unit. Remember, your goal here is to put some ideas on the table to help your students make sense of what they are seeing and doing, not to tell them what they should have learned.

Adding Information to the Summary Table (10 minutes)

1. Give students 5 minutes to decide what to add to the Summary Table at the end of the Student Handout.

2. Have one student from each group share what the group figured out, how they know (their evidence for what they figured out), and how this information will help them explain the anchoring phenomenon.

3. Once each group has shared, ask the entire class to decide what should be added to each column of the class Summary Table for Task 5. Help students reach consensus about what to add to the Summary Table. Only add an idea to the Summary Table if everyone in the class agrees with that idea.

Back Pocket Questions

As students work in groups, it is important to engage with each group to help press and extend students' thinking around the ideas at play in this task. Following are some example questions you might ask:

1. Helping students get started: What part of the root are you looking at? What structures do you see? What might be the function of those structures? How do the different cells look in comparison with each other? Are some bigger than others? Why might some of the cells look different from others? How many differences can you find when comparing all these cells?

2. Pressing further: This root would have kept growing over time if it had not been killed in the process of making this slide. What do you think would have happened to these structures if this root kept growing?

3. Following up: What makes you think that? Can you say more?

Filling Out the Summary Table

Table 1.12 includes examples of the responses students may come up with when they fill out the Summary Table. This is provided here only as a behind-the-scenes roadmap and is not meant to be shared with students.

Table 1.12. Example Summary Table for Task 5

What we learned	How it helps us explain the phenomenon
In plants, as in all multicellular organisms, individual cells grow and then divide via a process called mitosis, thereby allowing the organism to grow.	The cells inside General Sherman increase in number as the tree grows. The cells do not simply grow in size. Each new cell that is created has the same DNA as the original cell so the whole plant has the information it needs to function and keep growing.

Teacher Notes

Hints for Implementing This Task

- The videos can be viewed as a whole class for discussion if necessary.

Possible Extensions

- Although this task has not focused on the specific phases of mitosis and cell division, you might provide students with additional details based on the level of the course.

Student Handout

TASK 5. HOW DO PLANTS GROW?

Introduction

General Sherman, like all trees, started out as a small seed. The embryo inside the seed began increasing in size many years ago once it germinated. After a few days of growing, the plant was large enough to be called a seedling. General Sherman gets the energy it needs to grow by breaking the bonds of sugar and oxygen molecules and forming new compounds through a chemical process called *cellular respiration*. This is the same process that animals use to get the energy they need to move around, find food, grow, and reproduce. Unlike animals, however, plants do not need to eat something that contains sugar to get the sugar they need for cellular respiration. Instead, plants can produce the sugar they need through a chemical process called *photosynthesis*. The process of photosynthesis converts light energy into stored chemical energy by converting carbon dioxide found in the air plus water into sugar and oxygen molecules (which are byproducts of the process). The sugar molecules can then be combined with other sugar molecules to create larger carbohydrates, such as sucrose and starch molecules. These molecules are used to store chemical energy in plant tissues. Photosynthesis is the main way that energy from the Sun is captured and stored in the bodies of plants. The processes of photosynthesis and cellular respiration therefore provide most of the energy that General Sherman needed to grow from a seed to a huge tree.

You now need to figure out what happens inside a plant as it grows and increases in size in order to explain how General Sherman got so big. One way to learn more about what happens inside a plant as it grows is to look at the cells inside a specific structure that we know is growing, such as the tip of a root. The roots of plants are a system that has a specific structure and performs an essential function. Roots, like other plant systems, consist of a system of specialized cells. These specialized cells enable the plant to absorb water and nutrients from the soil, and they allow some plants, such as carrots, onions, and yams, to store large amounts of starch for later use. The roots of a plant also anchor it in the ground. Roots are constantly growing as they spread out through the soil to provide more water, nutrients, and stability for a plant as it increases in size.

Initial Ideas

Before beginning this task, take a few minutes to think about the structures you might find inside the tip of root, what the function of those structures might be, and how those structures change as the root increases in size. On the following page, create a model of your initial ideas by drawing and labeling the structures you think are found inside the tip of a root and how each of these structures changes as it grows. Be sure to include a description of what you think each structure does in your labels.

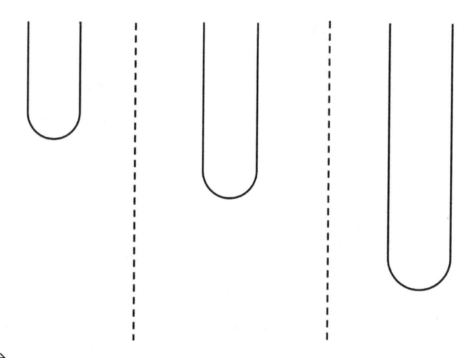

Time →

Your Task

Your task is to use what you know about the *growth and development of organisms* and the relationship between *structure and function* in living things to *develop a model* to explain what happens inside the root of a plant as it increases in size over time. To accomplish this task, you will use a microscope to examine the cells in the tip of an onion root. This is a good part of a plant to examine to learn more about how plants increase in size, because the root of a plant is always growing. However, the root was killed when the slide was made. As a result, the slide you will look at allows you to see what is happening inside the root as it grows only at one moment in time. You will not be able to see how the structures inside the root change over time as it grows. You will therefore also watch several time-lapse videos of what happens to the cells found inside an animal as it grows. Although these videos do not show plant cells, they might be helpful as you develop your model. Be sure to keep track of anything you observe or measure as you work in the Observations and Measurements section of your handout.

Available Materials

You and your group may use any of the following materials during this task:

- Microscope
- Prepared slide of an onion root tip
- A computer or tablet with an internet connection

Safety Precautions

Follow all normal lab safety rules. In addition, be sure to take the following safety precautions:

- Use only GFI-protected circuits when using electrical equipment, and keep away from water sources to prevent shock.
- Wash hands with soap and water when the activity is completed.

Observations and Measurements

Use the space below to keep a record of what you observe or measure as you work with the slide showing the cells in the tip of an onion root.

Activity

Watch the following videos to see what happens to the cells of animal as it grows:

- *www.youtube.com/watch?v=N97cgUqV0Cg*
- *www.youtube.com/watch?v=2LBpF71GZbk*
- *www.youtube.com/watch?v=ZeW8HaCUtOQ*

Keep a record of what you notice as you watch these videos in the space below.

Revise Your Model

Draw and label how you think a root increases in size over time, based on what you observed when you looked through a microscope at the structures that make up the tip of an onion root and what you saw in the videos of animal cells dividing. Be sure to include a description of what you think happens to the different structures in your labels.

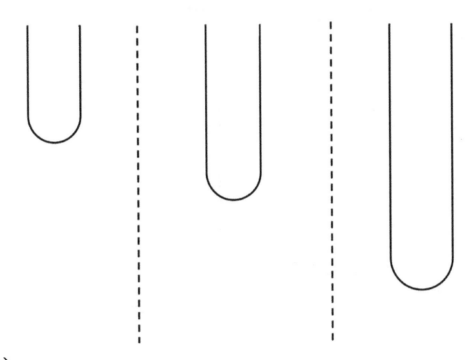

Time →

Some Useful Ideas From My Teacher

You can keep track of useful ideas from your teacher in the space below.

What We Figured Out

Now that you have completed this task, take a few minutes to fill out the Summary Table below with the other students in your group. This table will help you keep track of what you figured out during the task. You will then have an opportunity to share your ideas as we fill out our class Summary Table.

Summary Table

What we learned	How we know

How it helps us explain the phenomenon

National Science Teaching Association

Teacher Notes

Task 6. Where Does All That Matter Come From?

Purpose

This task gives students an opportunity to learn more about how plants obtain the matter they need to create new cells. To accomplish this task, students analyze and interpret data from two different experiments, one a now-classic experiment conducted in 1936 by scientist Paul Kramer (Kramer 1937), the other a fictional experiment with data we generated based on common lab activities performed by high school biology students. Performing this task helps students figure out the source of all the matter that makes up the cells, tissues, and different systems of a tree.

Important Life Science Content

All life-forms are made up of large, complex molecules called biological macromolecules or bio-molecules, which are produced by the cells of living organisms and are responsible for life. Bio-molecules contain carbon, meaning they are organic compounds. There are four major types of biomolecules: carbohydrates, lipids, proteins, and nucleic acids.

Carbohydrates are made from sugar and can be either simple, with only one or two rings of sugar, or complex, with three or more rings of sugar. Simple carbohydrates, such as glucose, are broken down and absorbed quickly and are a readily available energy source for living organisms. Complex carbohydrates provide organisms with long-term energy because they are stored in the cells for later use.

Lipids include fats, phospholipids, steroids, and waxes. They are made of chains of carbon atoms bonded to each other as well as to hydrogen atoms. Lipids are a more efficient way for organisms to store energy than carbohydrates because they contain a larger number of carbon and hydrogen atoms.

Proteins are made up of chains of amino acids held together by peptide bonds. The chains twist and fold into certain shapes, and the shape determines what the protein is supposed to do. Proteins carry out most cell and tissue functions, including producing other molecules, speeding up chemical reactions, and breaking down waste.

Nucleic acids are made up of carbon, hydrogen, oxygen, nitrogen, and phosphorus atoms. They are involved in creating the proteins needed to build and maintain cells, tissues, and organisms.

Scientific Ideas That Are Important to Think About During This Task

- Most of the matter that makes up a plant comes from the air.
- The sugar molecules formed through photosynthesis contain carbon, hydrogen, and oxygen. The hydrocarbon backbones of sugars are used to make amino acids and other carbon-based molecules that can be assembled into larger molecules (such as cellulose, proteins, or DNA) and then used to create new cells.

Timeline

Approximately two class periods.

Materials and Preparation

The items needed for this investigation are listed in Table 1.13.

Table 1.13. Required Materials for Task 6

Item	Quantity
Calculator	1 per group
2 × 3 foot dry-erase whiteboard	1 per group
Dry-erase pens	1 set per group
Student Handout	1 per student

Safety Precautions

This task does not require any specific safety precautions.

Procedure

This lesson plan is only a suggestion. It is included here to illustrate how you can facilitate student thinking during this task. We encourage you to modify this lesson plan by asking different questions, using different examples, and providing different scaffolds as appropriate to better meet the needs of students in your class.

Introduction of the Task (10 minutes)

1. Have students sit in groups of three or four.
2. Give each student one copy of the Student Handout for Task 6.
3. Read the Introduction and Initial Ideas sections of the handout aloud, having students follow along. It is a good idea to show a picture of General Sherman and seeds from a giant sequoia at this point in the lesson so students have the anchoring phenomenon of the unit in mind as they start this task.
4. Ask students to draw and label on the handout how they think plants obtain the matter that is needed to create new cells. Remind them to include labels that describe what is happening in their pictures.
5. Give students an opportunity to share their ideas with the others in their group.
6. Ask one student from each group to share their ideas with the entire class.

Analyzing and Interpreting Data and Then Arguing From Evidence (60 minutes)

1. Read the section of the handout called Your Task aloud to students.
2. Go over the two experiments and show them the results of each one. Be sure to ask questions about how they might make sense of the data. Don't be afraid to offer some

suggestions, but do not require that every group do everything the same way. Give them a choice.

3. Give students about 20 minutes to analyze the data. As students work, move from group to group and check in with them. It's important to ask them questions that will help them connect what they are doing to the goal of the task and the anchoring phenomenon. (See the Back Pocket Questions for Task 6 on page 106 for some suggestions.)

4. Go over the section called Your Argument with students. Tell them that they need to create an argument on a whiteboard to share with others what they figured out and their evidence in support of what they figured out. They also need to write down things they are unsure about on the board so their classmates can provide them with feedback.

5. Give the students 15 minutes to create an argument on a whiteboard.

6. As students work, again move from group to group and check in with them, asking questions such as those in the Back Pocket Questions on page 106.

7. Give students about 10 minutes to share, evaluate, and revise their arguments. We recommend using a modified gallery walk format for this part of the task. In this format, one or two members of the group (the "presenters") stay at their workstation to share their group's ideas, while the other group members (the "travelers") go to different groups one at a time to listen to and critique the arguments developed by their classmates. We recommend that the travelers visit at least three different workstations during the argumentation session, and that the presenters keep a record of the critiques made by their classmates and any suggestions for improvement. The travelers should also be encouraged to keep a record of good ideas or potential ways to improve their own arguments as they visit different groups.

8. Give students about 10 minutes to meet with their original groups so they can discuss what they learned by interacting with individuals from other groups and can revise their initial arguments. This process can begin with the presenters sharing the critiques and suggestions for improvement that they heard during the argumentation session. The students who visited other groups can then share their ideas for making the arguments better based on what they observed and discussed at other stations.

9. Have students clean up their workstations and then move on to the next part of the lesson.

Putting Ideas on the Table (10 minutes)

1. Give an interactive lecture to introduce some ideas that students might find useful for making sense of what they did. We recommend, as a minimum, introducing or reminding students of the following ideas:

   - As matter flows through different organizational levels of living systems, such as plants, chemical elements are recombined in different ways to form different products.

- The process of photosynthesis converts carbon dioxide plus water into sugars (and oxygen, which is released as a waste product). These sugar molecules contain carbon, hydrogen, and oxygen.
- The hydrocarbon backbones of sugar molecules can be used to make amino acids and other carbon-based molecules.
- The carbon-based molecules can be combined into even larger molecules, such as cellulose, proteins, and DNA, which can be used to create new cells.
- Cells are combined into tissues, tissues work together to form organs, and groups of organs that have specific functions based on their structure make up systems. Systems carry out the major functions of an organism.

2. Encourage students to keep a record of the ideas they find important in the section called Some Useful Ideas From My Teacher on the Student Handout so they can refer to them later in the unit. Remember, your goal here is to put some ideas on the table to help your students make sense of what they are seeing and doing, not to tell them what they should have learned.

Adding Information to the Summary Table (10 minutes)

1. Give students 5 minutes to decide what to add to the Summary Table at the end of the Student Handout.

2. Have one student from each group share what the group figured out, how they know (their evidence for what they figured out), and how this information will help them explain the anchoring phenomenon.

3. Once each group has shared, ask the entire class to decide what should be added to each column of the class Summary Table for Task 6. Help students reach consensus about what to add to the Summary Table. Only add an idea to the Summary Table if everyone in the class agrees with that idea.

Back Pocket Questions

As students work in groups, it is important to engage with each group to help press and extend students' thinking around the ideas at play in this task. Following are some example questions you might ask:

1. Helping students get started: How would you know if matter is moving into or out of a plant? Why was it important to include a cup without a plant in it in the second experiment?

2. Pressing further: What types of calculations could you make to help identify any patterns in these data? What types of mathematical representations could you create to help identify any patterns in these data?

3. Following up: What makes you think that? Can you say more?

Filling Out the Summary Table

Table 1.14 includes examples of the responses students may come up with when they fill out the Summary Table. This is provided here only as a behind-the-scenes roadmap and is not meant to be shared with students.

Table 1.14. Example Summary Table for Task 6

What we learned	How it helps us explain the phenomenon
Most of the matter that makes up a plant comes from the air. The sugar molecules formed through photosynthesis contain carbon, hydrogen, and oxygen. The hydrocarbon backbones of sugars are used to make amino acids and other carbon-based molecules that can be assembled into larger molecules (such as cellulose, proteins, or DNA) and then used to create new cells.	The sugar that General Sherman makes through the process of photosynthesis can also be used to make amino acids and other carbon-based molecules. These carbon-based molecules can then be combined into even larger molecules, such as cellulose, proteins, and DNA. The larger molecules are used to form new cells, which are combined into tissues. Tissues work together to form organs, which are organized into systems that carry out the functions of the tree.

Hints for Implementing This Task

- Some students may need assistance with interpreting the data table and producing the graph.

Possible Extensions

- To make the question of where the mass comes from more visceral, bring a piece of wood to class and pass it around so students can feel the weight.

Reference

Kramer, P. J. 1937. The relation between rate of transpiration and rate of absorption of water in plants. *American Journal of Botany* 24 (1): 10–15.

Student Handout

TASK 6. WHERE DOES ALL THAT MATTER COME FROM?

Introduction

The previous tasks showed us that plants, like General Sherman, grow by creating more cells. Individual cells inside the different tissues of a plant grow and then divide via mitosis, which allows the organism to increase in size by adding more cells and to replace dead or damaged cells in the tissue. In this task, you will learn more about how plants get the matter they need to create more cells and increase in size.

Remember that cells are composed of carbohydrates, lipids, proteins, and nucleic acids. Carbohydrates are made up of carbon, hydrogen, and oxygen atoms and include sugars and starches. Lipids are also made up of carbon, hydrogen, and oxygen atoms but have a different molecular structure. They include fats and oils, waxes, and steroids. Proteins consist of hundreds or thousands of amino acids, which are made up of carbon, hydrogen, oxygen, nitrogen, and sulfur atoms. The sequence of amino acids determines the function of the protein. Nucleic acids are made up of carbon, hydrogen, oxygen, nitrogen, and phosphorus atoms. Living organisms use nucleic acids in creating the proteins needed to build and maintain functioning cells, tissues, and organisms.

In this task, you will figure out the source of all the matter that makes up the cells, tissues, organs, and all the different systems of a tree. To accomplish this task, you will analyze data that were collected during two different experiments. The people who carried out these experiments measured the mass of plants, soil, and water over time to track the movement of matter into a plant. The change in mass of a system, such a plant, can be used to track how much matter moves into or out of that system, because all matter has mass and the total number of atoms that make up a material does not change when that material changes states or goes through a chemical reaction.

Initial Ideas

Before beginning this task, take a few minutes to think about where a plant might get the matter it needs to create new cells. In the space below, draw and label how you think a plant gets the matter it needs to make new cells. Be sure to include labels that describe what is happening.

Your Task

Your task is to use what you know about the *transfer of matter into and out of systems* to *analyze and interpret data* from two different experiments that were carried out to determine the source of all the matter that makes up the cells, tissues, and different systems of a tree. Then craft an argument to share what you have figured out with your classmates and explain how you know what you figured out is valid or acceptable.

Experiment 1: Does a Plant's Mass Come From Water?

One source of matter for plants could be water. A water molecule, after all, is made up of two hydrogen atoms and one oxygen atom, and hydrogen and oxygen atoms are found in all four of the biomolecules that make up cells. To test the idea that plants get the matter they need to create new cells from water, a scientist named Paul Kramer carried out a now-classic experiment in 1936 to determine how much water different types of plants absorb and lose over the course of a day. Kramer used the following procedure:

1. He filled 12 containers with 50 grams of soil, then added water until the moisture of the soil was about 18%.

2. He planted green ash plants in containers 1–3, pine plants in containers 4–6, cactus plants in containers 7–9, and sunflower plants in containers 10–12.

3. To each container, he added an autoirrigator, which would supply water to the soil as the plant absorbed water from the soil.

4. Next, he covered the tops of the pots with a double layer of oilcloth and packed cotton around the stems to prevent water evaporation from the soil.

5. Then, every 4 hours, he determined and recorded the amount of water absorbed and lost by the plants. The amount of water absorbed by the plant in a container was determined by measuring the amount of water required to fill the reservoir in the autoirrigator to a given mark in milliliters. The amount of water lost by the plant in a container was determined by weighing the entire system.

Tables SH6.1 and SH6.2 (p. 110) include some of the data that Paul Kramer collected during this experiment.

Table SH6.1. Amount of Water Absorbed by Plants During Experiment I

Container	Plant type	Water *absorbed* by the plant over time (mm)					
		8 a.m.	12 p.m.	4 p.m.	8 p.m.	12 a.m.	4 a.m.
1	Ash	2	8	20	15	4	3
2	Ash	3	9	19	15	5	2
3	Ash	1	8	21	16	4	4
4	Pine	5	40	42	20	5	5
5	Pine	7	36	39	16	4	4
6	Pine	3	44	45	24	6	5
7	Cactus	2	1	2	6	5	4
8	Cactus	2	1	3	5	5	3
9	Cactus	3	2	2	6	4	3
10	Sunflower	5	28	35	18	8	6
11	Sunflower	3	25	30	17	6	5
12	Sunflower	7	31	38	19	9	7

Source: The authors generated these data based on graphs in Kramer (1937). The data are intended for educational use only.

Table SH6.2. Amount of Water Lost by Plants During Experiment I

Container	Plant type	Water *lost* by the plant over time (mm)					
		8 a.m.	12 p.m.	4 p.m.	8 p.m.	12 a.m.	4 a.m.
1	Ash	2	28	26	3	2	2
2	Ash	3	29	25	4	2	3
3	Ash	3	28	27	3	1	1
4	Pine	8	50	49	18	4	2
5	Pine	9	48	48	17	3	2
6	Pine	7	53	52	20	5	2
7	Cactus	1	2	5	5	3	3
8	Cactus	1	1	4	4	2	2
9	Cactus	1	2	5	4	3	2
10	Sunflower	5	28	35	18	8	6
11	Sunflower	6	25	34	14	7	5
12	Sunflower	5	30	37	23	9	6

Source: The authors generated these data based on graphs in Kramer (1937). The data are intended for educational use only.

Use the space below to create a graph to help make sense of the results of this experiment. You may need to make some calculations as part of your analysis.

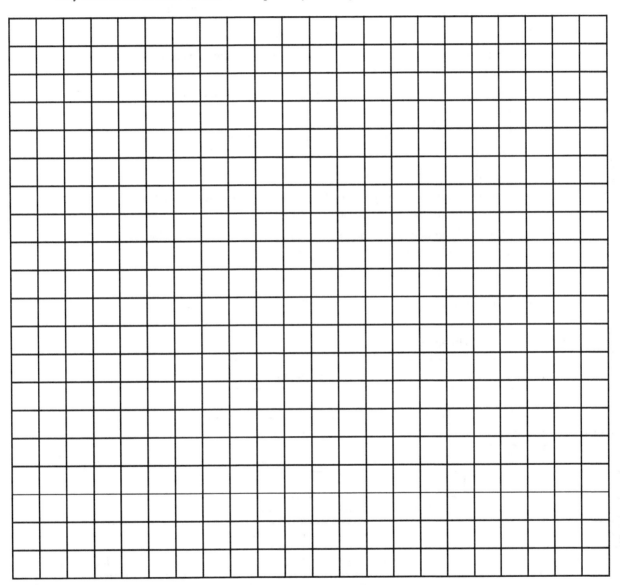

Experiment 2: Does a Plant's Mass Come From the Soil?

Another source of matter for plants could be the soil. The typical composition of soil is generally about 45% minerals, 5% organic matter, 25% water, and 25% air. Some of the most common minerals found in soil are iron, potassium, magnesium, calcium, and sulfur. Most organic matter in soil is tissue from dead plants. Plant residues contain 60–90% water. The remaining dry matter consists of carbon, oxygen, hydrogen, and small amounts of sulfur, nitrogen, calcium, phosphorus, potassium, and magnesium. All these elements are found in at least one of the four different bio-molecules that make up cells. To test the idea that plants get the matter they need to create new cells from the soil, a group of students carried out the following experiment:

1. First, they dried out 500 grams of soil in an oven.
2. Then they rinsed as much soil away as possible from the roots of four different plants that had been growing for weeks.
3. They weighed five different cups, then put 70–90 grams of dry soil into each cup.
4. Next, they weighed the four plants.
5. In four of the cups, they planted the four different plants, leaving the fifth cup with only dry soil.
6. Then they watered the plants daily.
7. After 14 days, they carefully removed the plants from the cups and rinsed their roots over filter paper to recover as much soil as possible.
8. They dried and weighed all the soil from each cup, including any soil washed away from the plants and collected on the filter paper.
9. Finally, they weighed each of the plants.

Table SH6.3 includes the data the students collected during this experiment.

Table SH6.3. Data from Experiment 2

| Cup number | Mass (grams) | | | | | |
| | Day 1 | | | Day 14 | | |
	Plant	Soil	Cup	Plant	Soil	Cup
1	23.7	84.9	9.2	26.9	84.6	9.2
2	26.8	76.5	8.4	29.7	75.9	8.4
3	24.6	77.8	8.8	28.1	77.1	8.8
4	28.2	86.1	9.2	33.9	85.3	9.2
5	—	80.6	8.4	—	80.6	8.4

Source: The authors generated these fictional data based on common student lab activities.

Use the space below to create a graph to help make sense of the results of this experiment. You may need to make some calculations as part of your analysis.

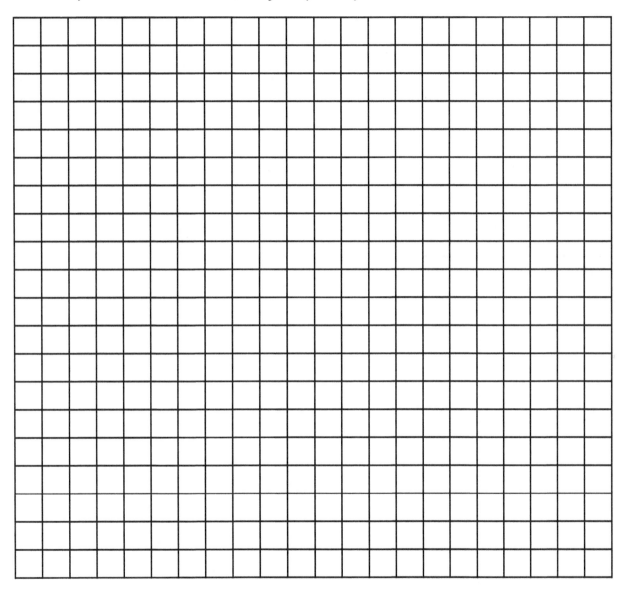

Your Argument

Develop an argument on a whiteboard. It should include these three components:

1. *What you figured out:* State a claim about how or why something happens.

2. *Your evidence:* Support your claim with an analysis of the available data and an interpretation of what the analysis means.

3. *Things you are unsure about:* List anything your group would like to learn more about in order to develop a more complete claim or to provide stronger evidence for your claim.

The Guiding Question:	
Our Claim:	
Our Evidence:	Our Justification of the Evidence:

Share your argument with your classmates. Be sure to ask them how to make your draft argument better. Keep track of their suggestions in the space below.

Ways to IMPROVE our argument ...

Some Useful Ideas From My Teacher

You can keep track of useful ideas from your teacher in the space below.

What We Figured Out

Now that you have completed this task, take a few minutes to fill out the Summary Table below with the other students in your group. This table will help you keep track of what you figured out during the task. You will then have an opportunity to share your ideas as we fill out our class Summary Table.

Summary Table

What we learned	How we know

How it helps us explain the phenomenon

Reference

Kramer, P. J. 1937. The relation between rate of transpiration and rate of absorption of water in plants. *American Journal of Botany* *24* (1): 10–15.

Teacher Notes

Task 7. What's Inside a Tree?

Purpose

This task gives students an opportunity to learn more about how structures of a tree's roots, trunk, branches, and leaves help it grow. To accomplish this task, students analyze and interpret data that were collected during two different experiments and watch three videos about the internal structures of plant organs. Performing this task helps students figure out where photosynthesis takes place in General Sherman and how water, minerals, carbon dioxide, and other molecules transfer into, within, and out of this giant tree. *Note:* Both experiments in this task are based on common lab activities performed by high school biology students. We generated fictional data sets to give students an opportunity to participate in SEP 4 (analyzing and interpreting data) and SEP 5 (using mathematics and computational thinking). These data are consistent in all respects with what we would expect students to observe in the described experiments.

Important Life Science Content

Plants, and specifically trees, have common structures, implying common functions. The crown, which consists of the leaves and branches at the top of a tree, filters dust and other particles from the air, helps cool the air by providing shade, and reduces the impact of raindrops on the soil below. The crown of every tree is unique, with some being broad and far-reaching and others being very tall and narrow.

A tree's roots absorb water (along with nutrients dissolved in the water) from the soil, store sugar, and anchor the tree upright in the ground. Root structure differs among species of trees, some with large and extensive below-ground structures and others with aerial roots seen above the ground.

The trunk, or stem, supports the crown and give the tree its shape and strength. The trunk consists of layers of tissue that form a network that runs between the roots and leaves and acts as a "central plumbing" system for the tree. This system includes two types of transport tissue, xylem and phloem, with a layer of cambium in between. The xylem, or sapwood, makes up the youngest layers of wood. The xylem brings water and nutrients up from the roots through the inside of the trunk to the leaves and other parts of the tree. Xylem cells only carry water up but not down. As a tree grows, older xylem cells in the center of the tree become inactive and die, forming heartwood, which helps support the tree. Heartwood is dark in color, as it is full of stored sugar, dyes, and oils.

The cambium is a thin layer of growing tissue that produces new cells that become xylem, phloem, or more cambium. Every growing season, a tree's cambium adds a new layer of xylem cells to its trunk, producing a visible growth ring. This is what makes the trunk, branches, and roots grow larger in diameter. The phloem, or inner bark, is found between the cambium and the outer bark and acts as a food supply line by carrying sap (sugar and nutrients dissolved in water) from the leaves to the rest of the tree. Phloem cells allow flow in both directions. The trunk, branches, and twigs of the tree are all covered with bark, which protects the tree from insects, disease, storms, and extreme temperatures. The bark originates from phloem cells that have worn out, died, and were shed outward.

A leaf is also a system that is made up different components. Leaves consist of three types of tissue: the epidermis, an outer protective layer; the mesophyll, which is rich in chloroplasts and found inside the leaf; and the vascular tissue, which includes the xylem and phloem and allows waters, minerals, and carbon-based molecules to move into or out of the leaf.

All cells in a tree are genetically identical, meaning they all have the same DNA and therefore instructions to become any part of the tree and any cell type. Based on environmental chemical signals, cells can determine what type of cell they need to become and use their common DNA instructions to create the proteins that make the different parts of the tree unique. This process is known as cell differentiation and occurs in almost every multicellular body. The chemical signal from the environment is received by a receptor on the outside of a cell, which induces a signal transduction pathway directed at the DNA in the nucleus to initiate a specific cellular response to start making proteins to become a certain type of cell. In humans, stems cells are known for their ability to differentiate. In trees, the cells of the cambium differentiate based on signals to become phloem or xylem cells. Cells near the crown differentiate to become leaf cells. It is through cell differentiation that there can be so many visually different and functionally unique portions of a tree and other multicellular beings.

Cellulose is a biomolecule that helps keep plants, such as General Sherman, stiff, tall, and strong. Cellulose is a long-chain polymer of glucose molecules. A polymer is a long and repeating chain of the same molecule bound together. The cell wall of plant cells is primarily made up of cellulose. A glucose molecule is composed of carbon, hydrogen, and oxygen atoms. Looking at the periodic table shows that the mass of glucose (and therefore cellulose) is primarily from atoms of carbon and oxygen, since the mass of hydrogen is merely 1. The next task will make explicit that the source of the carbon and oxygen atoms is carbon dioxide from the atmosphere.

Scientific Ideas That Are Important to Think About During This Task

- Photosynthesis mainly takes place in the leaves of trees.
- Leaves are made up of specialized cells that help them perform the essential functions of life.
- Plants have vascular tissues that transport water from the ground to the leaves and sugar and other carbon-based molecules from the leaves to the rest of the plant.
- Leaves are also needed to move water through a plant.

Timeline

Approximately two class periods.

Materials and Preparation

The items needed for this investigation are listed in Table 1.15.

Table 1.15. Required Materials for Task 7

Item	Quantity
Calculator	1 per group
2 × 3 foot dry-erase whiteboard	1 per group
Dry-erase pens	1 set per group
Computer or tablet with internet connection	1 per group
Student Handout	1 per student

Safety Precautions

This task does not require any specific safety precautions.

Procedure

This lesson plan is only a suggestion. It is included here to illustrate how you can facilitate student thinking during this task. We encourage you to modify this lesson plan by asking different questions, using different examples, and providing different scaffolds as appropriate to better meet the needs of students in your class.

Introduction of the Task (10 minutes)

1. Have students sit in groups of three or four.

2. Give each student one copy of the Student Handout for Task 7.

3. Read the Introduction and Initial Ideas sections of the handout aloud, having students follow along. It is a good idea to show a picture of General Sherman and seeds from a giant sequoia at this point in the lesson so that students have the anchoring phenomenon of the unit in mind as they start this task.

4. Ask students to draw and label on the handout their ideas about the structures and functions of the roots, trunk, branches, and leaves of a tree. Remind them to be sure to include labels that describe what is happening in their pictures.

5. Give students an opportunity to share their ideas with the others in their group.

6. Ask one student from each group to share their ideas with the entire class.

Analyzing and Interpreting Data and Then Arguing From Evidence (60 minutes)

1. Read the section of the handout called Your Task aloud to students.

2. Go over the two experiments and show students the results of each one. Be sure to ask questions about how they might make sense of the data. Don't be afraid to offer some suggestions, but do not require that every group do everything the same way. Give them a choice.

3. Give students about 20 minutes to analyze the data. As students work, move from group to group and check in with them. It's important to ask them questions that will help them connect what they are doing to the goal of the task and the anchoring phenomenon. (See the Back Pocket Questions for Task 7 on pages 121–122 for some suggestions.)

4. Go over the section called Your Argument with students. Tell them that they need to create an argument on a whiteboard to share with others what they figured out and their evidence in support of what they figured out. They also need to write down things they are unsure about on the board so their classmates can provide them with feedback.

5. Give students 15 minutes to create an argument on a whiteboard.

6. As students work, again move from group to group and check in with them, asking questions such as those in the Back Pocket Questions on pages 121–122.

7. Give students about 10 minutes to share, evaluate, and revise their arguments. We recommend using a modified gallery walk format for this part of the task. In this format, one or two members of the group (the "presenters") stay at their workstation to share their group's ideas, while the other group members (the "travelers") go to different groups one at a time to listen to and critique the arguments developed by their classmates. We recommend that the travelers visit at least three different workstations during the argumentation session, and that the presenters keep a record of the critiques made by their classmates and any suggestions for improvement. The travelers should also be encouraged to keep a record of good ideas or potential ways to improve their own arguments as they visit different groups.

8. Give students about 10 minutes to meet with their original groups so they can discuss what they learned by interacting with individuals from other groups and can revise their initial arguments. This process can begin with the presenters sharing the critiques and suggestions for improvement they heard during the argumentation session. The students who visited other groups can then share their ideas for making the arguments better based on what they observed and discussed at other stations.

9. Have students clean up their workstations and then move on to the next part of the lesson.

Putting Ideas on the Table (35 minutes)

1. Give students about 15 minutes to watch the three videos showing the internal structures of leaves, the trunk, and branches of a tree, as well as an overview of the transport system of plants.

2. Remind students to keep a record of the ideas that they find important as they watch the videos in the space on the Student Handout.

3. Give an interactive lecture (about 20 minutes maximum) to introduce some ideas that students might find useful for making sense of what they did. We recommend, as a minimum, introducing or reminding students of the following ideas:

- Trees, like General Sherman, have a hierarchical structural organization, in which any one system is made up of numerous parts, and that system is one of the many components that make up the next level of organization.
- The cells of plants can each be viewed as an individual system that is made up of components including a cell wall, mitochondria, and chloroplasts, which together allow the cell to carry out specific functions that are needed for survival, such as photosynthesis and respiration.
- An organ, such as a leaf, is also a system that is made up different components. These components are called tissues.
- Leaves consist of three different types of tissue: the epidermis, an outer protective layer; the mesophyll, which is rich in chloroplasts and found inside the leaf; and the vascular tissue, which includes the xylem and phloem and allows waters, minerals, and carbon-based molecules to move into or out of the leaf.

4. Encourage students to keep a record of the ideas they find important in the section called Some Useful Ideas From My Teacher on the Student Handout so they can refer to them later in the unit. Remember, your goal here is to put some ideas on the table to help your students make sense of what they are seeing and doing, not to tell them what they should have learned.

Adding Information to the Summary Table (10 minutes)

1. Give students 5 minutes to decide what to add to the Summary Table at the end of the Student Handout.
2. Have one student from each group share what the group figured out, how they know (their evidence for what they figured out), and how this information will help them explain the anchoring phenomenon.
3. Once each group has shared, ask the entire class to decide what should be added to each column of the class Summary Table for Task 7. Help students reach consensus about what to add to the Summary Table. Only add an idea to the Summary Table if everyone in the class agrees with that idea.

Back Pocket Questions

As students work in groups, it is important to engage with each group to help press and extend students' thinking around the ideas at play in this task. Following are some example questions you might ask:

1. Helping students get started: How would you know if photosynthesis is happening inside a plant or not? How would you know if carbon dioxide is moving into or out of a plant? Which containers do you need to compare to figure out what is happening?

2. Pressing further: What types of calculations could you make to help identify any patterns in these data? What types of mathematical representations could you create to help identify any patterns in these data?

3. Following up: What makes you think that? Can you say more?

Filling Out the Summary Table

Table 1.16 includes examples of the responses students may come up with when they fill out the Summary Table. This is provided here only as a behind-the-scenes roadmap and is not meant to be shared with students.

Table 1.16. Example Summary Table for Task 7

What we learned	How it helps us explain the phenomenon
Photosynthesis takes place mainly in the leaves of trees. Leaves are made up of specialized cells that help them perform the essential functions of life. Plants have vascular tissues that transport water from the ground to the leaves and sugar and other carbon-based molecules from the leaves to the rest of the plant. Leaves are also needed to move water through a plant.	General Sherman gets the carbon dioxide it needs for photosynthesis through pores in its needles called stomata. Water enters the tree through the roots, moves up the trunk through the xylem, and then enters the needles. It also exits the leaves through the stomata.

Hints for Implementing This Task

• The videos can be viewed as a whole class for discussion if necessary.

Possible Extensions

• You can use the classic celery-in-colored-water experiment to demonstrate the transportation of water through xylem.

Student Handout

TASK 7. WHAT'S INSIDE A TREE?

Introduction

During the last task, you figured out that chemical elements are recombined in different ways to form different products as matter flows through the different organizational levels of a living system, such as General Sherman. The process of photosynthesis, for example, converts carbon dioxide molecules from the air and water molecules into sugar molecules and oxygen molecules. The sugar molecules created during photosynthesis contain carbon, hydrogen, and oxygen. A plant, such as General Sherman, can then use these sugar molecules to make other carbon-based molecules such as starch and cellulose. These molecules can be broken down, and the atoms are then combined with other elements, such as phosphorus, nitrogen, and sulfur, to make other molecules, including amino acids, lipids, and nucleic acids. The amino acids are combined to make proteins, and these proteins, along with other large carbon-based molecules, are used to create the various structures inside new cells.

In plants, just like in animals, cells that are found in the same location and have a similar structure and function combine to form a tissue. When different types of tissues work together to perform a unique function, they form an organ; organs work together to form an organ system. Plants, such as trees, have three distinct organ systems: the shoot system, the root system, and the reproductive system. The shoot system is composed of the leaves, petioles or branches, and the stem or trunk of the plant. The root system contains the roots of the plant. The reproductive system consists of flowers and fruits. The shoot system and reproductive system grow aboveground, and the root system grows underground.

You know that the process of photosynthesis takes place in the leaves of a plant because you analyzed and interpreted data from an experiment that measured carbohydrate formation in the large leaves of sunflower plants. However, many plants have small leaves and a very large root system, and others have a stem and no branches or leaves. Trees have a woody stem called a trunk and many different woody branches that have leaves on them. General Sherman, for example, is estimated to have a mass of about 4.2 million pounds, and the trunk, branches, and roots make up about 95% of its total mass. So what do these parts of the tree do? Do the leaves of plants have other functions in addition to being a site for photosynthesis?

Your goal in this task is to figure out the answers to these two questions. To accomplish this goal, you will have the opportunity to analyze and interpret data that were collected during two experiments: one that examined photosynthesis in different plant parts and one that examined the role of different plant parts in the absorption of water from the environment. These experiments were designed to help shed some light on the various functions of the roots, trunks, branches, and leaves of trees. You will also have a chance to watch some videos that will provide you with some additional information about these structures.

Initial Ideas

Before beginning this task, take a few minutes to think about the structures and functions of the roots, trunk, branches, and leaves of a tree. In the space that follows, draw and label what you think is inside these structures and how they function. Be sure to include labels that describe what is happening.

Your Task

Your task is to use what you know about the *transfer of matter into and out of systems* to *analyze and interpret data* from two different experiments about the functions of the roots, trunks, branches, and leaves of trees. Then you will watch three videos about the structures found inside a tree's leaves, trunk, and branches to *obtain some additional information* about them. After you have completed these two activities, you will need to craft an argument to share what you have figured out with your classmates and explain how you know what you figured out is valid or acceptable.

Experiment 1: Does Photosynthesis Take Place in Plant Organs Besides Leaves?

To test the idea that photosynthesis occurs throughout a plant and not just in the leaves, a group of students carried out the following experiment:

1. First, they took 100-g samples of tissue from different plant parts.

2. They placed each sample in a clear container of equal volume and sealed the containers.

3. Then they measured the carbon dioxide level in each container.

4. Next, they shone a light on the containers for a preestablished amount of time.

5. Finally, they again measured the carbon dioxide level in each container.

Table SH7.1 includes the data that the students collected during this experiment. Assume that all the conditions not listed in the table were identical in all 24 of the containers.

Table SH7.1. Data from Experiment I

| Container | Plant type | Plant part | Time in light (min.) | Carbon dioxide level (ppm) | |
				Begin	End
1	Myrtle	Leaf	60	400	386
2	Myrtle	Leaf	30	410	401
3	Myrtle	Leaf	45	415	402
4	Myrtle	Leaf	60	405	387
5	Oak	Leaf	60	402	391
6	Oak	Leaf	30	410	405
7	Oak	Leaf	45	412	404
8	Oak	Leaf	45	415	407
9	Myrtle	Root	60	400	736
10	Myrtle	Root	60	412	742
11	Myrtle	Root	30	405	567
12	Myrtle	Root	30	415	583
13	Oak	Root	45	420	654
14	Oak	Root	50	415	675
15	Oak	Root	60	418	730
16	Oak	Root	45	412	650
17	Myrtle	Stem	60	414	690
18	Myrtle	Stem	30	415	553
19	Myrtle	Stem	60	417	687
20	Myrtle	Stem	30	415	556
21	Oak	Trunk	50	404	644
22	Oak	Trunk	60	407	695
23	Oak	Trunk	45	410	612
24	Oak	Trunk	45	415	618

Source: The authors generated these fictional data based on common student lab activities.

Use the space below to create a graph to help make sense of the results of this experiment. You may need to make some calculations as part of your analysis.

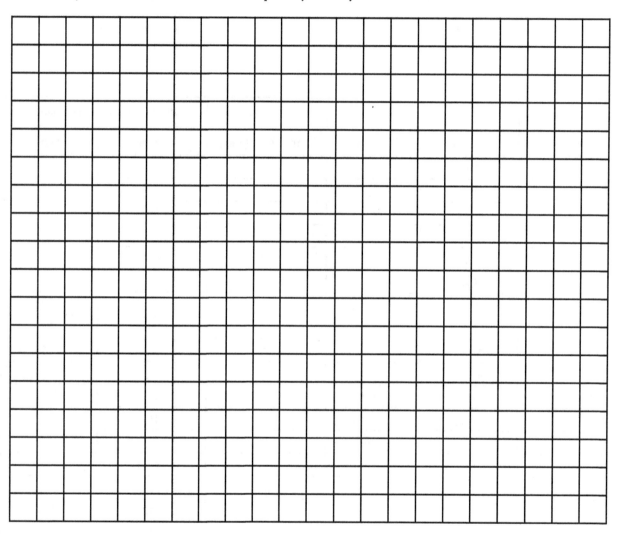

Experiment 2: Which Organs Are the Most Important for Getting the Water Plants Need for Photosynthesis?

Plants need water to survive. In fact, access to water is the factor that limits plant growth the most. Water is so important to plants because it is required for photosynthesis. Despite the important role water plays in the growth and survival of plants, most plants retain only about 5% of the water they absorb over time in the new cells and tissues they create as they grow. The remainder passes through plants directly into the atmosphere. To determine which structures plants use to get the water they need to survive and grow, a group of students carried out the following experiment:

1. First, they obtained 21 healthy plants with at least 10 leaves and a well-developed root system.
2. They removed all soil from the roots of each plant and then prepared the different plants as follows.
3. Plants 1–3: removed 100% of the leaves.
4. Plants 4–6: removed 100% of the roots.
5. Plants 7–9: cut the stems at the midpoint.
6. Plants 10–12: removed 100% of the leaves and 100% of the roots.
7. Plants 13–15: removed 50% of the leaves and 50% of the roots.
8. Plants 16–18: removed 50% of the roots.
9. Plants 19–21: nothing removed or cut.
10. After this, they weighed each plant.
11. They added about 20 ml of water to each of 21 graduated cylinders.
12. Next, they added one plant to each graduated cylinder.
13. They measured the exact water level in each graduated cylinder.
14. After waiting 20 minutes, they again measured the water level in each graduated cylinder.

Table SH7.2 (p. 128) includes the data that the students collected during this experiment. Assume that all the conditions not listed in the table were identical in all 21 of the containers. The experiment also included another graduated cylinder with exactly 20.0 ml of water in it and no plant as a control. The water level in that graduated cylinder did not change after 20 minutes.

Table SH7.2. Data From Experiment 2

Container	Modifications to plant			Mass of plant (g)	Water level (ml)	
	Roots	Stem	Leaves		Begin	End
1	No change	No change	100% removed	80.7	20.1	20.0
2	No change	No change	100% removed	82.4	19.9	19.7
3	No change	No change	100% removed	83.1	19.3	19.0
4	100% removed	No change	No change	80.3	19.3	14.7
5	100% removed	No change	No change	85.1	19.5	15.3
6	100% removed	No change	No change	86.4	19.8	15.2
7	No change	Cut at midpoint	No change	102.3	20.1	20.0
8	No change	Cut at midpoint	No change	100.4	19.9	19.7
9	No change	Cut at midpoint	No change	100.9	19.3	19.0
10	100% removed	No change	100% removed	50.4	19.3	19.2
11	100% removed	No change	100% removed	51.1	20.3	20.3
12	100% removed	No change	100% removed	50.2	19.4	19.3
13	50% removed	No change	50% removed	76.1	19.5	16.6
14	50% removed	No change	50% removed	78.9	19.8	17.2
15	50% removed	No change	50% removed	77.2	19.2	16.9
16	50% removed	No change	No change	90.2	19.5	14.7
17	50% removed	No change	No change	91.4	19.2	15.1
18	50% removed	No change	No change	91.7	21.1	15.3
19	No change	No change	No change	100.2	19.2	14.6
20	No change	No change	No change	100.5	19.7	15.2
21	No change	No change	No change	101.1	19.8	14.9

Source: The authors generated these fictional data based on common student lab activities.

Use the space below to create a graph to help make sense of the results of this experiment. You may need to make some calculations as part of your analysis.

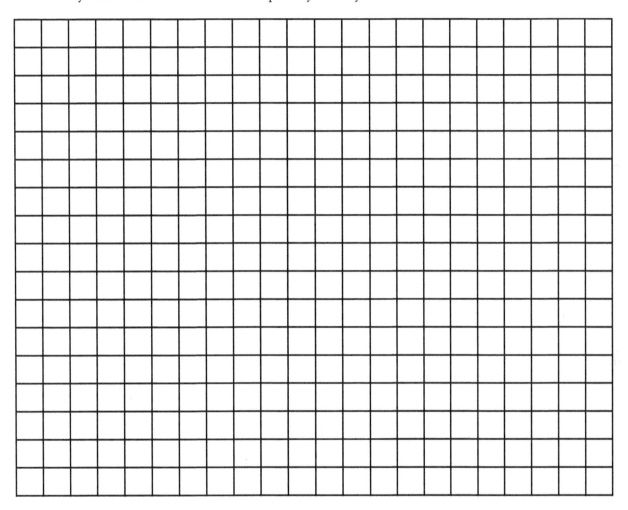

Your Argument

Develop an argument on a whiteboard. It should include these three components:

1. *What you figured out:* State a claim about how or why something happens.
2. *Your evidence:* Support your claim with an analysis of the available data and an interpretation of what the analysis means.

The Guiding Question:	
Our Claim:	
Our Evidence:	Our Justification of the Evidence:

3. *Things you are unsure about:* List anything your group would like to learn more about in order to develop a more complete claim or to provide stronger evidence for your claim.

Share your argument with your classmates. Be sure to ask them how to make your draft argument better. Keep track of their suggestions in the space below.

Ways to IMPROVE our argument ...

Activity

Watch the following videos to take a look inside the leaves, trunk, and branches of a tree. These videos will provide some additional information about the structures found inside:

- *www.youtube.com/watch?v=pwymX2LxnQs*
- *www.youtube.com/watch?v=9-dicqNoODg*
- *www.youtube.com/watch?v=R0wjTdBK77o*

Keep a record of any important ideas in the space below as you watch these videos.

Some Useful Ideas From My Teacher

You can keep track of useful ideas from your teacher in the space below.

National Science Teaching Association

What We Figured Out

Now that you have completed this task, take a few minutes to fill out the Summary Table below with the other students in your group. This table will help you keep track of what you figured out during the task. You will then have an opportunity to share your ideas as we fill out our class Summary Table.

Summary Table

What we learned	How we know

How it helps us explain the phenomenon

Teacher Notes

Task 8. Does the Environment Affect Plant Growth?

Purpose

This task gives students an opportunity to learn more about how environmental conditions affect plant growth. To accomplish this task, students analyze and interpret data collected by scientists during two different experiments and watch two videos about how tree rings can be used to study the growth of trees over time. Performing this task helps students figure out how changes in environmental conditions affect the growth of General Sherman over time.

Important Life Science Content

Plant growth is greatly affected by the environment. If any environmental factor is less than ideal, it limits a plant's growth. This is known as a limiting factor. For example, only plants adapted to low levels of water can live in deserts. Environmental factors that affect plant growth include light, temperature, water, and nutrition. It is important to understand how these factors affect plant growth and development.

Three principal characteristics of light affect plant growth: quantity, quality, and duration. Light quantity refers to the intensity, or concentration, of sunlight. It varies with the seasons. The maximum amount of light is present in summer and the minimum in winter. Up to a point, the more sunlight a plant receives, the greater its capacity for producing food via photosynthesis. Light quality refers to the color (wavelength) of light. Sunlight supplies the complete range of wavelengths and can be broken up by a prism into bands of red, orange, yellow, green, blue, indigo, and violet. Blue and red light, which plants absorb, have the greatest effect on plant growth. Blue light is responsible primarily for vegetative (leaf) growth. Red light, when combined with blue light, encourages flowering. Plants look green to us because they reflect, rather than absorb, green light. Finally, duration, or photoperiod, refers to the amount of time a plant is exposed to light.

Temperature influences most plant processes, including photosynthesis, transpiration, respiration, germination, and flowering. As temperature increases, up to a point, photosynthesis, transpiration, and respiration increase. When combined with daylength, temperature also affects the change from vegetative (leafy) to reproductive (flowering) growth. Depending on the situation and the specific plant, the effect of temperature can be to either speed up or slow down this transition.

Most growing plants contain about 90 percent water, which plays many roles in plants. It is a primary component in photosynthesis and respiration, a regulator of stomatal opening and closing, the source of pressure to move roots through the soil, and the medium in which most biochemical reactions take place. It is also responsible for turgor pressure in cells and for cooling leaves as it evaporates from leaf tissue during transpiration. Because of its important role in photosynthesis (among many other plant processes), water is an important limiting factor of photosynthesis.

Finally, plant nutrition refers to a plant's need for and use of basic chemical elements. Plants need 17 elements for normal growth. Three of them—carbon, hydrogen, and oxygen—are found in air and water. The rest are found in the soil. Six soil elements are called macronutrients because they are used in relatively large amounts by plants: nitrogen, potassium, magnesium, calcium, phosphorus, and sulfur. A lack of any of these chemical elements can be a limiting factor in plant growth.

Scientific Ideas That Are Important to Think About During This Task

- Photosynthesis is most efficient with brighter light and warmer temperatures.
- If it gets too hot or too cold, chlorophyll breaks down and photosynthesis stops.
- Trees do not grow as much during droughts or when it gets cold.

Timeline

Approximately two class periods.

Materials and Preparation

The items needed for this investigation are listed in Table 1.17.

Table 1.17. Required Materials for Task 8

Item	Quantity
Calculator	1 per group
Computer or tablet with internet connection	1 per group
Student Handout	1 per student

Safety Precautions

This task does not require any specific safety precautions.

Procedure

This lesson plan is only a suggestion. It is included here to illustrate how you can facilitate student thinking during this task. We encourage you to modify this lesson plan by asking different questions, using different examples, and providing different scaffolds as appropriate to better meet the needs of students in your class.

Introduction of the Task (10 minutes)

1. Have students sit in groups of three or four.
2. Give each student one copy of the Student Handout for Task 8.
3. Read the Introduction and Initial Ideas sections of the handout aloud, having students follow along. It is a good idea to show a picture of General Sherman and seeds from a giant sequoia at this point in the lesson so that students have the anchoring phenomenon of the unit in mind as they start this task.

4. Ask students to draw and label on the handout their ideas about how trees might grow under different enviromental conditions, such as during the winter, spring, and summer. Remind them to include labels that describe what is happening in their pictures.

5. Give students an opportunity to share their ideas with the others in their group.

6. Ask one student from each group to share their ideas with the entire class.

Analyzing and Interpreting Data and Using Mathematics and Computational Thinking (35 minutes)

1. Read the section of the handout called Your Task aloud to students.

2. Go over the two experiments and show students the results of each one. Be sure to ask questions about how they might make sense of the data. Don't be afraid to offer some suggestions, but do not require that every group do everything the same way. Give them a choice.

3. Give students about 25 minutes to analyze the data. As students work, move from group to group and check in with them. It's important to ask them questions that will help them connect what they are doing to the goal of the task and the anchoring phenomenon. (See the Back Pocket Questions for Task 8 for some suggestions.)

Putting Ideas on the Table (35 minutes)

1. Give students about 15 minutes to watch the two videos about how tree rings can be used to study the growth of trees over time.

2. Remind students to keep a record of the ideas they find important as they watch the videos in the space on the Student Handout.

3. Give an interactive lecture (about 20 minutes maximum) to introduce some ideas students might find useful for making sense of what they did. We recommend, as a minimum, introducing or reminding students of the following ideas:

   - Locations near the poles experience large seasonal differences in high and low daily temperatures, duration of daylight, and amount of precipitation, whereas locations near the equator do not experience seasonal changes in temperature and duration of daylight (although these locations often experience a rainy season and a dry season).

   - If it gets too hot or too cold, over 140°F (60°C) or under 68°F (20°C), the chlorophyll in the cells of leaves will break down and photosynthesis will stop.

4. Encourage students to keep a record of the ideas they find important in the section called Some Useful Ideas From My Teacher on the Student Handout so they can refer to them later in the unit. Remember, your goal here is to put some ideas on the table to help your students make sense of what they are seeing and doing, not to tell them what they should have learned.

Adding Information to the Summary Table (10 minutes)

1. Give students 5 minutes to decide what to add to the Summary Table at the end of the Student Handout.

2. Have one student from each group share what the group figured out, how they know (their evidence for what they figured out), and how this information will help them explain the anchoring phenomenon.

3. Once each group has shared, ask the entire class to decide what should be added to each column of the class Summary Table for Task 8. Help students reach consensus about what to add to the Summary Table. Only add an idea to the Summary Table if everyone in the class agrees with that idea.

Back Pocket Questions

As students work in groups, it is important to engage with each group to help press and extend students' thinking around the ideas at play in this task. Following are some example questions you might ask:

1. Helping students get started: How would you know whether photosynthesis is happening inside a plant? How would you know if carbon dioxide is moving into or out of a plant? Which containers do you need to compare to figure out what is happening?

2. Pressing further: What types of calculations could you make to help identify any patterns in these data? What types of mathematical representations could you create to help identify any patterns in these data?

3. Following up: What makes you think that? Can you say more?

Filling Out the Summary Table

Table 1.18 includes examples of the responses students may come up with when they fill out the Summary Table. This is provided here only as a behind-the-scenes roadmap and is not meant to be shared with students.

Table 1.18. Example Summary Table for Task 8

What we learned	How it helps us explain the phenomenon
Photosynthesis is most efficient with brighter light and warmer temperatures. If it gets too hot or too cold, chlorophyll breaks down and photosynthesis stops. Trees do not grow as much during droughts or colder weather.	General Sherman needs constant access to bright light, water, and warm temperatures in order to grow. When it gets too cold or there is a drought, it does not grow as well.

Hints for Implementing This Task

- Some students may need assistance with interpreting the data table and producing the graph.

Possible Extensions

- To demonstrate how tree ring information is collected, you can purchase a small handheld tree borer through a forestry supply company.

References

Brun, W. A., and R. L. Cooper. 1967. Effects of light intensity and carbon dioxide concentration on photosynthetic rate of soybean. *Crop Science* 7 (5): 451–454.

Fock, H., K. Klug, and D. T. Canvin. 1979. Effect of carbon dioxide and temperature on photosynthetic CO2 uptake and photorespiratory CO2 evolution in sunflower leaves. *Planta* 145 (3): 219–223.

Student Handout

TASK 8. DOES THE ENVIRONMENT AFFECT PLANT GROWTH?

Introduction

Trees, like General Sherman, have a hierarchical structural organization, in which any one system is made up of numerous parts, and that system is one of the many components that make up the next level of organization. The cells of plants, for example, can each be viewed as an individual system that is made up of components including a cell wall, mitochondria, and chloroplasts, which together allow the cell to carry out specific functions that are needed for survival, such as photosynthesis and respiration. An organ, such as a leaf, is also a system that is made up different components, called tissues. Leaves consist of three different types of tissue: the epidermis, an outer protective layer; the mesophyll, which is rich in chloroplasts and found inside the leaf; and the vascular tissue, which includes the xylem and phloem and allows waters, minerals, and carbon-based molecules to move into or out of the leaf.

During the previous task, you figured out that photosynthesis takes place mainly in the leaves of trees. You also figured out that water moves through plants because of a process called transpiration. In transpiration, water is absorbed from the soil by the roots and transported as a liquid to the leaves through the xylem. This water then escapes through small pores in the leaves, called stomata, as a gas. Most plants retain only about 5% of the water they absorb from the ground as they grow. The rest of the water passes through plants directly into the atmosphere as a result of transpiration. The amount of water lost by a plant due to transpiration can be quite high. For example, a single corn plant can lose close to 200 liters of water in three months, and some large trees found in rainforests can lose almost 1,200 liters of water in a single day.

Not all plants on Earth, however, have access to plenty of water. Some plants are found in areas that receive more than 100 inches of rain a year, while others are found in areas that get less than 10 inches. The amount of sunlight also differs in different locations. Take Phoenix, Arizona, and Seattle, Washington, as examples. Phoenix has 3,872 hours of sunlight each year, but Seattle has only 2,170 hours. The temperature also differs throughout the year in different locations. For example, in Phoenix, the average high temperature is 106°F (41°C) in July and 68°F (20°C) in January, while in Seattle, the average high is 72°F (22°C) in July and 47°F (8°C) in January. Some locations can get even colder in the winter. For example, in Bismarck, North Dakota, the average high is only 23°F (–5°C) in January.

Your goal in this task is to figure out whether and how changes in the environment over time affect tree growth. To accomplish this goal, you will have an opportunity to first analyze and interpret data that scientists collected during an experiment designed to examine how different environmental conditions affect photosynthesis. You will then have a chance to watch some videos that will provide you with some additional information about how scientists study tree growth.

Initial Ideas

Before beginning this task, take a few minutes to think about how trees might grow under different environmental conditions. For example, the temperature, duration of daylight, and amount of rainfall at a location often change with the seasons. In the space that follows, draw and label how you think these changes in environmental conditions affect the growth of a tree. Be sure to include labels that describe what is happening.

Your Task

Your task is to use what you know about the *transfer of matter into and out of systems* to *analyze and interpret data* from two experiments that examined how light intensity and temperature affected photosynthesis. Then you will watch two videos about the way scientists study tree growth over time.

Experiment 1: How Does Light Intensity Affect Photosynthesis?

Two scientists, W. A. Brun and R. L. Cooper, examined how changing light intensity affects photosynthesis in the late 1960s. They carried out the following experiment:

1. They grew 12 soybean plants for three weeks in the same conditions.
2. Then they placed an intact leaf from one of the plants in a sealed container (without removing the leaf from the plant).
3. They measured the carbon dioxide level in the container.
4. Next, they shone a light with a predetermined intensity on the leaf for 60 minutes.
5. Then they again measured the carbon dioxide level in the container.
6. They repeated the procedure with the remaining 11 plants.

Table SH8.1 includes some of the data the scientists collected during this experiment. Assume that all the conditions not listed in the table were identical in containers 1–12.

Table SH8.1. Data From Experiment I

Container	Plant type	Light intensity (lux)	Temperature (°C)	Time (min.)	Carbon dioxide level (ppm)	
					Begin	End
1	Soybean	5,380	27	60	402	392
2	Soybean	5,380	27	60	412	401
3	Soybean	5,380	27	60	409	400
4	Soybean	5,380	27	60	415	404
5	Soybean	21,330	27	60	403	378
6	Soybean	21,330	27	60	408	384
7	Soybean	21,330	27	60	416	393
8	Soybean	21,330	27	60	415	389
9	Soybean	43,000	27	60	404	377
10	Soybean	43,000	27	60	411	383
11	Soybean	43,000	27	60	409	383
12	Soybean	43,000	27	60	414	386

Source: The authors generated these data based on graphs in Brun and Cooper (1967). The data are intended for educational use only.

Use the space below to create a graph to help make sense of the results of this experiment. You may need to make some calculations as part of your analysis.

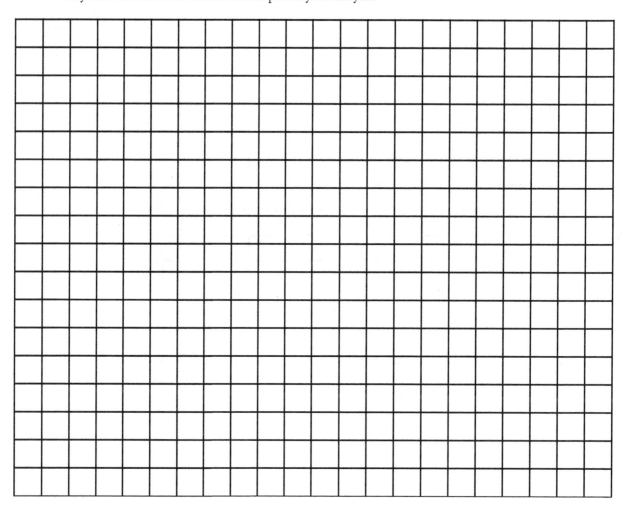

Experiment 2: How Does Temperature Affect Photosynthesis?

About 10 years later, scientists Heinrich Fock, Klaus Klug, and David Canvin examined how temperature affects photosynthesis. Their experiment was as follows:

1. First, they transplanted 12 sunflower seedlings that were 40 days old, all grown under the same conditions, into individual containers filled with garden soil.

2. They placed each sample in a clear container of equal volume and sealed the containers.

3. Then they changed the temperature in the containers.

4. They measured the carbon dioxide level in each container.

5. Next, they shone a light on the plants in each container for 60 minutes.

6. Finally, they again measured the carbon dioxide level in each container.

Table SH8.2 includes some of the data that the scientists collected during this experiment. Assume that all the conditions not listed in the table were identical in containers 1–12.

Table SH8.2. Data From Experiment 2

Container	Plant type	Light intensity (lux)	Temperature (°C)	Time (min.)	Carbon dioxide level (ppm)	
					Begin	End
1	Sunflower	6,330	16	60	415	403
2	Sunflower	6,330	16	60	414	403
3	Sunflower	6,330	16	60	418	405
4	Sunflower	6,330	16	60	420	409
5	Sunflower	6,330	24	60	419	403
6	Sunflower	6,330	24	60	415	398
7	Sunflower	6,330	24	60	417	401
8	Sunflower	6,330	24	60	413	396
9	Sunflower	6,330	34	60	409	385
10	Sunflower	6,330	34	60	412	387
11	Sunflower	6,330	34	60	415	392
12	Sunflower	6,330	34	60	411	386

Source: The authors generated these data based on graphs in Fock, Klug, and Canvin (1979). The data are intended for educational use only.

Use the space below to create a graph to help make sense of the results of this experiment. You may need to make some calculations as part of your analysis.

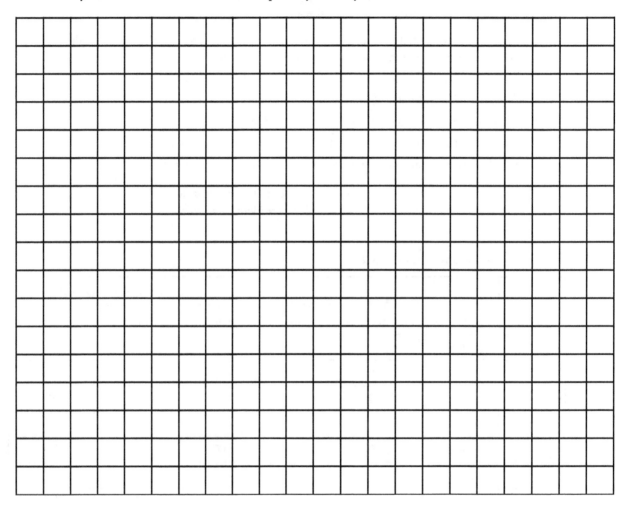

National Science Teaching Association

Activity

Watch the following videos to see how scientists study tree growth over time. These videos will provide some additional information about how changes in the environment affect the growth of a tree and branches of a tree:

- *www.youtube.com/watch?v=mbdur2TjTbk*
- *www.youtube.com/watch?v=xmZO7aRgcW4*

Keep a record of any important ideas in the space below as you watch these videos.

Some Useful Ideas From My Teacher

You can keep track of useful ideas from your teacher in the space below.

What We Figured Out

Now that you have completed this task, take a few minutes to fill out the Summary Table below with the other students in your group. This table will help you keep track of what you figured out during the task. You will then have an opportunity to share your ideas as we fill out our class Summary Table.

Summary Table

What we learned	How we know

How it helps us explain the phenomenon

References

Brun, W. A., and R. L. Cooper. 1967. Effects of light intensity and carbon dioxide concentration on photosynthetic rate of soybean. *Crop Science* 7 (5): 451–54.

Fock, H., K. Klug, and D. T. Canvin. 1979. Effect of carbon dioxide and temperature on photosynthetic CO_2 uptake and photorespiratory CO_2 evolution in sunflower leaves. *Planta* 145 (3): 219–23.

National Science Teaching Association

Teacher Notes

Task 9. How Do Trees Survive in Different Climates?

Purpose

This task gives students an opportunity to learn more about the adaptations of trees and other plants that enable them to survive in different climates. To accomplish this task, students first watch several videos about different types of plant adaptations. They then look up climate data for a town near Sequoia National Park where General Sherman is located. Performing this task helps students figure out how General Sherman has survived long enough to grow so large.

Important Life Science Content

Like all living things, giant sequoias are well adapted to their environment. General Sherman grows within Sequoia National Park in California. Coniferous forests, which include giant sequoia groves, dominate the middle elevations of the park from approximately 4,000 to 9,000 feet. At these elevations, precipitation typically occurs during October to mid-May, when the forests receive about 45 inches annually on average. Snow is common in winter when the temperatures are cooler.

Conifers like giant sequoias have a number of adaptations to survive in the winter months. First, the tree's bark acts as its first line of defense against the cold. The outer bark protects the tree from disease, insects, storms, and extreme temperatures. It is full of air spaces and acts as insulation for the tree, similar to the insulation in the walls of your home.

Coniferous trees have different leaf adaptations to survive in the winter. Their needles are long and thin, giving them a small surface area. Less surface area means they have fewer stomata from which to lose water, and the stomata are not on the surface but are deep within the needles. This creates a pocket of still air just inside the needle, which results in less transpiration than moving air. Narrow leaves also help keep snow from building up and breaking branches. Finally, a conifer needle has a thick, waxy cuticle, the outermost layer of a leaf. The wax helps prevent water loss. This is important in the summer months as well, as both the stomata (and accompanying guard cells) limit water loss due to high temperatures.

Beneath a tree's bark is a system of transport tissues that act like pipelines to move water, sugar, and other nutrients up and down the tree. This system consists of xylem, which move water and nutrients upward from the tree's roots to its leaves, and phloem, which move sugars down from the leaves to the rest of the tree. The fluids that move up and down the tree are known as sap.

A lot of water moves up and down the tree through the xylem and phloem, and if this water freezes, it can destroy the tree's cells. This is because ice crystals are sharp and can tear through cell walls. So just like in your home, it is important that a tree's "pipes" do not freeze. One way the tree works to prevent ice damage is by controlling where ice forms. Ice has to form around something. Outdoors, it usually forms around things like bits of dust. In plants, ice can form around certain types of molecules called ice nucleators. Trees produce proteins that act as ice nucleators and send them between cells. As ice begins to form, it draws liquid water out of the tree's cells. As more and more water is pulled out of the cells, the sap left inside the cells contains more and more sugars and becomes thick and syrupy. It now has a freezing point lower than the environment, making it more difficult to freeze. In addition to the nucleators, trees also produce antifreeze proteins, which help prevent ice crystals from forming in cold temperatures.

Scientific Ideas That Are Important to Think About During This Task

- The leaves, branches, and roots of plants are adapted to maximize photosynthesis and minimize water loss in hot climates and damage from water freezing in cold climates.
- Sequoia National Park has cold, wet winters and hot, dry summers. The area gets about 45 inches of rain annually, and snow is common during the winter months.

Timeline

Approximately one class period.

Materials and Preparation

The items needed for this investigation are listed in Table 1.19.

Table 1.19. Required Materials for Task 9

Item	Quantity
Computer or tablet with internet connection	1 per group
Student Handout	1 per student

Safety Precautions

This task does not require any specific safety precautions.

Procedure

This lesson plan is only a suggestion. It is included here to illustrate how you can facilitate student thinking during this task. We encourage you to modify this lesson plan by asking different questions, using different examples, and providing different scaffolds as appropriate to better meet the needs of students in your class.

Introduction of the Task (10 minutes)

1. Have students sit in groups of three or four.
2. Give each student one copy of the Student Handout for Task 9.
3. Read the Introduction and Initial Ideas sections of the handout aloud, having students follow along. It is a good idea to show a picture of General Sherman and seeds from a giant sequoia at this point in the lesson so that students have the anchoring phenomenon of the unit in mind as they start this task.

4. Ask students to draw and label on the handout their ideas about how trees that are found in a hot and wet climate look different from trees that are found in a cold and dry climate. Remind them to include labels that describe what is happening in their pictures.

5. Give students an opportunity to share their ideas with the others in their group.

6. Ask one student from each group to share their ideas with the entire class.

Obtaining and Evaluating Information (45 minutes)

1. Read the section of the handout called Your Task aloud to students.

2. Give students about 20 minutes to watch the four videos about plant adaptations to different climates.

3. Remind students to keep a record of the ideas they find important as they watch the videos in the space on the Student Handout.

4. Give students about 25 minutes to use the U.S. Climate Data website to look up information about the climate in Sequoia National Forest.

5. Be sure to ask questions about how students might make sense of the data. Don't be afraid to offer some suggestions, but do not require that every group do everything the same way. Give them a choice.

6. As students work, move from group to group and check in with them. It's important to ask them questions that will help them connect what they are doing to the goal of the task and the anchoring phenomenon. (See the Back Pocket Questions for Task 9 on page 150 for some suggestions.)

Putting Ideas on the Table (15 minutes)

1. Give an interactive lecture to introduce some ideas students might find useful for making sense of what they did. We recommend, as a minimum, introducing or reminding students of the following ideas:

   • In any particular environment, some kinds of organisms survive well, some survive less well, and some cannot survive at all.

   • Adaptations are anatomical, behavioral, or physiological traits that make an organism well suited to survive and reproduce in a specific environment.

2. Encourage students to keep a record of the ideas they find important in the section called Some Useful Ideas From My Teacher on the Student Handout so they can refer to them later in the unit. Remember, your goal here is to put some ideas on the table to help your students make sense of what they are seeing and doing, not to tell them what they should have learned.

Adding Information to the Summary Table (10 minutes)

1. Give students 5 minutes to decide what to add to the Summary Table at the end of the Student Handout.

2. Have one student from each group share what the group figured out, how they know (their evidence for what they figured out), and how this information will help them explain the anchoring phenomenon.

3. Once each group has shared, ask the entire class to decide what should be added to each column of the class Summary Table for Task 9. Help students reach consensus about what to add to the Summary Table. Only add an idea to the Summary Table if everyone in the class agrees with that idea.

Back Pocket Questions

As students work in groups, it is important to engage with each group to help press and extend students' thinking around the ideas at play in this task. Following are some example questions you might ask:

1. Helping students get started: What types of information about the climate do you think are the most and least important? Is it more important to look at the atmospheric conditions during specific months or to look at change over time?

2. Pressing further: Do you think the weather conditions where General Sherman is located are relatively stable from season to season, or does the weather change a lot during the year?

3. Following up: What makes you think that? Can you say more?

Filling Out the Summary Table

Table 1.20 includes examples of the responses students may come up with when they fill out the Summary Table. This is provided here only as a behind-the-scenes roadmap and is not meant to be shared with students.

Table 1.20. Example Summary Table for Task 9

What we learned	How it helps us explain the phenomenon
The leaves, branches, and roots of plants are adapted to maximize photosynthesis and minimize water loss in hot climates and damage from water freezing in cold climates. Sequoia National Park has cold, wet winters and hot, dry summers. The area gets about 45 inches of rain annually, and snow is common during the winter months.	The leaves, branches, and roots of General Sherman are adapted to minimize water loss, maximize water absorption, and prevent cell damage during freezing temperatures. These adaptations helped this tree survive and grow for thousands of years.

Hints for Implementing This Task

- You can illustrate interesting adaptations by bringing a diverse set of plants to the classroom, including succulents.

Possible Extensions

- Students can download data to analyze historical climate fluctuations in Sequoia National Park.

Student Handout

TASK 9. HOW DO TREES SURVIVE IN DIFFERENT CLIMATES?

Introduction

In the previous task, you learned about the factors that limit photosynthesis, such as light, temperature, and water availability. This is important because all living things, including General Sherman, need to be adapted to their environments. The important limiting factors for photosynthesis and plant growth are different in different types of climates. For General Sherman, this means being adapted to the climate in Sequoia National Park in California.

During the previous task, you figured out that changes in environmental conditions affect the process of photosynthesis. Photosynthesis is most efficient with brighter light and warmer temperatures. However, if it gets too hot or too cold, the chlorophyll in the cells of leaves will break down and photosynthesis will stop. As a result, high temperatures during the summer and low temperatures during the winter can slow down or completely stop the process of photosynthesis from happening in plants.

You also know that plants need access to plenty of water for photosynthesis. Plants are able to absorb the water they need from the ground with their roots, and then transport it to their leaves using transport tissues called xylem. However, plants also lose a great deal of water due to transpiration. Water loss can be a huge problem for plants when there is not much water available in the environment around them. The water inside a plant can also freeze if it gets too cold outside. When water freezes inside a cell, it expands and destroys the cell. If too many cells are destroyed inside the plant, the plant will die.

So how do plants survive in locations that are very dry, very hot, or very cold? We know that all plants, including General Sherman, have a hierarchical structural organization in which each system is made up of numerous parts, and that each system is one of the many components that make up the next level of organization. You also know that the cells, tissues, and organs that make up these systems are highly specialized and perform specific functions that are essential for life. It is therefore important to examine the ways the tissues and organs of plants are structured in order to figure out how plants are able to live in different locations that have very different climates.

Your goal in this task is to figure out how the structures of General Sherman have enabled it to survive in a California climate for so long. To accomplish this goal, you will first watch some videos about the types of adaptations plants have that allow them to survive in different climates, which will provide you with some information about how plant structures differ in plants that live in different climates. You will then learn more about the climate where General Sherman is located so you can begin to generate some ideas about how this tree was able to survive so long and grow so big over time.

Initial Ideas

Before beginning this task, take a few minutes to think about how the structure of the roots, trunk, branches, and leaves of a tree, like General Sherman, might enable it to survive in different types of climates. In the space that follows, draw and label what you think the roots, trunk, branches, and leaves of a tree that is found in a hot and wet climate and a tree that is found in a cold and dry climate might look like. Be sure to include labels that describe what is happening.

Your Task

Your task is to use what you know about the *structure and function* to figure out how *a tree like General Sherman* is able to live so long in the climate of Sequoia National Park in California. You will first watch four videos about the structures of plants found in different climates. Then you will do some research to learn more about the climate where General Sherman is located.

Activity 1

Watch the following videos to see the tree structures found in different climates. These videos will provide you with some additional information about the leaves, roots, and trunks of trees and how these plant structures differ in different climates:

- *www.britannica.com/video/152187/overview-leaf-structure-functions-plant*
- *www.youtube.com/watch?v=d260CmZoxj8*
- *www.youtube.com/watch?v=H9MV5CgPgIQ*
- *www.youtube.com/watch?v=zu6p3kdfqwg*

Keep a record of any important ideas in the space below as you watch these videos.

Activity 2

Do some research to learn what the climate is like in Sequoia National Park in California, where General Sherman is located. The town of Three Rivers is near this park. Visit the U.S. Climate Data website at *www.usclimatedata.com* to find out more about the climate in Three Rivers and Sequoia National Park. Keep a record of any useful information you find in the space below.

Some Useful Ideas From My Teacher

You can keep track of useful ideas from your teacher in the space below.

What We Figured Out

Now that you have completed this task, take a few minutes to fill out the Summary Table below with the other students in your group. This table will help you keep track of what you figured out during the task. You will then have an opportunity to share your ideas as we fill out our class Summary Table.

Summary Table

What we learned	How we know

How it helps us explain the phenomenon

Building Consensus Stage Summary

In the building consensus stage of an MBI unit (Figure 1.13), students finalize their group models, compare and contrast those models as a class as they work to reach a consensus, and co-construct a Gotta Have Checklist of the ideas and evidence that should be part of their final evidence-based explanations. Before beginning this stage, review the relevant sections in Chapters 1 and 2 for specifics. See Figures 1.3 and 1.4 (pp. 28–29) for examples of a Gotta Have Checklist and a final model for this unit. We have provided a PowerPoint template for this stage, which can be downloaded from the book's Extras page at *www.nsta.org/mbi-biology*.

Figure 1.13. Building Consensus Stage Summary

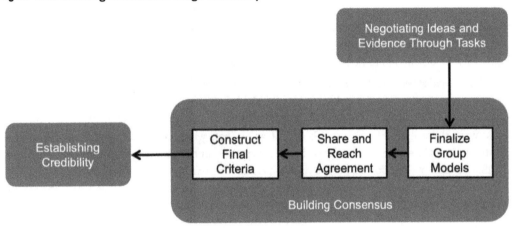

Finalize Group Models

Finalizing the models requires groups to review their previous models, decide what needs to be revised based on new ideas and understandings from the previous set of tasks, and redraw the models so they can be more easily shared and used to build consensus in a whole-class setting. This usually takes about 30 minutes. We often scaffold this process by asking students to review the completed Summary Table and talk about what should be added to, removed from, or changed in their previous models. They should think about what new ideas can help them move toward a truly causal model that includes not only *what* happened but *why* it happened. This requires that the mechanism at play be visible and well explained. We push the groups to make sure that items in their models are labeled, that any unseen components are made visible, and that the important ideas that surfaced throughout the unit are explicitly used in the models.

Share and Reach Agreement

We think that a share-out session is needed here so the groups can learn from each other and begin to build consensus across the models. There are a number of ways to do this, including gallery walks. As the goal is to build consensus as a whole class, we usually opt to facilitate a share-out session in which each group comes to the front of the class, displays its model, and talks it through. We prompt students to ask questions, and the class works hard to compare and contrast ideas across the models.

Construct a List of Final Criteria

The goal of our last public record, the Gotta Have Checklist, is to have the class negotiate about which of the main ideas and evidence should be part of a complete and scientifically defensible explanation of the phenomenon. Students will use this as a scaffold for writing their final evidence-based explanations in the last stage of the unit. There are a number of ways to facilitate this discussion and creation of the public record. We like to prompt groups to create a bulleted list of the three to five most important ideas that they think they need. We then ask for examples, press the whole class to make sure they understand how the idea fits in, and ask for consensus before writing it on the final checklist public record. This process can take about 15 to 20 minutes and most often goes quite smoothly by this point in the unit. However, this is also a time to make sure important ideas are included and to bring them up if necessary. Though this is uncommon, we may also provide some just-in-time instruction to tie up any loose ends in students' understandings from the unit.

The checklist becomes less useful to students if everything that comes up in this discussion is automatically written on the board. Ideas can generally be combined into five to seven bulleted points. We then go back and lead a discussion about the evidence we have for each of the bulleted points and write those alongside. This step is crucial, as students will need to coordinate the science ideas with evidence in the written evidence-based explanation in the next stage. At times, we have been unhappy with our checklist for some reason; perhaps the writing is not as legible or the points are not as clearly articulated as we would like. In such cases, we created a clean version that evening and asked the students the next day to ensure that it still represented all the ideas from the original poster.

Establishing Credibility Stage Summary

In the establishing credibility stage (Figure 1.14), students construct arguments in the form of evidence-based explanations of the phenomenon, which allows them to argue their ideas in writing. This stage also includes peer review and revision to establish credibility for their ideas. Before beginning this stage, review the relevant sections in Chapters 1 and 2 for specifics. We have provided a PowerPoint template to assist with this stage, which can be downloaded from the book's Extras page at *www.nsta.org/mbi-biology*.

Figure 1.14. Establishing Credibility Stage Summary

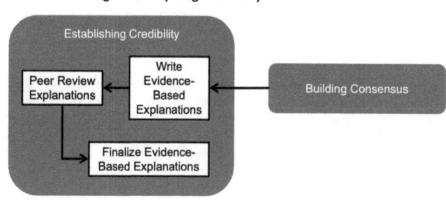

Write Evidence-Based Explanations

Up to this point, students have worked as part of a group to construct their evidence-based explanations of the phenomenon. Now, they work individually to write their final explanations. While they are on their own, they have a number of important public records such as the Summary Table and Gotta Have Checklist to use as references. Beyond these scaffolds, you may choose to provide additional support for students who may struggle with writing. For instance, we often have the first sentence written out on small slips of paper that we can hand to students to help build momentum. Some teachers choose to have student groups work together to create a group outline of their writing before students work individually. Your school may have writing support structures in place to help students get started. One thing we have found is that teachers are most often pleasantly surprised by their students' individual evidence-based explanations.

Conduct Peer Review of Explanations

Once students have finished writing their individual evidence-based explanations, it is important that they receive peer feedback before turning in their final product. Have students use the MBI Explanation Peer-Review Guide in Appendix B to provide feedback to each other. Beyond helping students improve their written explanations, this step reinforces the idea that in science, credibility comes not from an external standard but through negotiation with our peers.

Finalize Evidence-Based Explanations

After students have been able to review their peer feedback, and perhaps meet with their peers to discuss the feedback, they should use the critique to finalize their evidence-based explanations.

Evaluation

Once students have handed in their written explanations, you can use the MBI Explanation Peer-Review Guide in Appendix B to evaluate them. Besides clarity of communication, the guide focuses on whether a student's final product explains the phenomenon, fits with the evidence gathered throughout the unit, and builds on the important science ideas at the center of the unit.

Unit 2. Ecosystems

Interactions, Energy, and Dynamics (LS2): Whale Plumes Change the Atmosphere

Unit Summary

This unit on whale plumes incorporates the *Next Generation Science Standards* (*NGSS*) three-dimensional learning strategies through interactive high cognitive tasks as students develop models to construct explanations of a real-world phenomenon. The unit engages students with a discussion about whaling and introduces whale fecal plumes as a key factor in the anchoring phenomenon. The issue of whaling combined with the visual of whale plumes in the water creates a highly engaging phenomenon. This unit takes students through the disciplinary core idea (DCI) Life Science 2 (LS2) performance expectations covering the concept of ecosystems: interactions, energy, and dynamics. More specifically, students explore how whale plumes provide nutrients for an oceanic food web. This involves food webs and trophic cascades, especially in relation to energy flow. Subsequently, students learn about the role that whale plumes play in nitrogen and iron availability in the open oceans and how this affects phytoplankton growth (LS2.A). They then learn about the role that phytoplankton, historically sustained by nitrogen and iron from whale plumes, have played in environmental carbon fixation (LS2.B). In the end, students make connections between whaling and the elimination of a significant carbon reservoir (LS2.C). The crosscutting concept of energy and matter is important throughout the unit.

Table 2.1 provides an outline for this unit, describing the purpose and the major tasks and products for each MBI stage, as well as an approximate timeline.

Table 2.1. MBI Outline and Timeline

MBI stage	Purpose	Major tasks and products	Timing
Eliciting ideas about the phenomenon	To introduce the phenomenon, eliciting students' initial ideas to explain the phenomenon, and begin constructing group models.	Initial ideas public record; initial group models	1–2 days
Negotiating ideas and evidence through tasks	To provide opportunities for students to make sense of the phenomenon through purposeful tasks and discussions.	Individual task products; revised group models	9–10 days
Building consensus	For groups to come to an agreement about the essential aspects of the explanation.	Final group models	1 day
Establishing credibility	For each individual student to write an evidence-based explanation.	Final evidence-based explanations	1–2 days
		Total	12–15 days

Anchoring Phenomenon

The ocean has been an important cultural and economic resource for the island country of Japan for thousands of years. Fish constitutes a large portion of the Japanese diet, and Japan consumes millions of tons of fish each year. In recent decades, however, fisheries depletion has become a global issue. Between 1995 and 2011, the total catch in Japan decreased from 6 million to 3.8 million tons. Consequently, Japan decided to increase whaling, reasoning that whales eat fish and krill and were at least somewhat responsible for the decline in fish populations, either directly or indirectly through competition. However, with the increase of whaling, the fish populations began to crash. Scientists set out to better understand why the fish populations declined. In doing so, they discovered that not only did whales influence fish populations, but also their removal contributed to climate change. Students are oriented to whale plumes early in the unit as a potential important factor leading to these disturbances in the ecosystem.

Driving Question

Why did whaling lead to a collapse in fish populations and contribute to climate change?

Target Explanation

Following is an example evidence-based explanation that could be expected at the end of the unit. We consider this an exemplar final explanation at this grade level. This is included here to help support you, the teacher, in responsively supporting students in negotiating similar explanations by the end of the unit. It is not something to be shared with students but is only provided as a behind-the-scenes roadmap for you to consider before and throughout the unit to guide your instruction.

> Humans have been hunting whales for thousands of years. This has had far-reaching effects on the ecosystem. Recently, the Japanese government claimed that the hunting of whales would help the fishing industry by reducing competition for fish and krill at the bottom of the food chain. They reasoned that this would lead to an increase in the commercial fish populations. But the hunting of whales and the subsequent decline in whale populations had the opposite effect, and the fish and krill populations dramatically decreased. The decline in fish and krill populations can be attributed to many factors, but among the most important is the decline in the whale populations and the subsequent decrease in whale fecal plumes, which provide essential nutrients (e.g., nitrogen and iron) for phytoplankton.
>
> We learned that whales release large fecal plumes of nutrients near the surface of the ocean, and these can serve as a much-needed source of nutrients for phytoplankton. Phytoplankton live near the surface of the ocean because, as producers, they need sunlight to carry out the process of photosynthesis. In the Limiting Factors for Phytoplankton task, we learned that phytoplankton use nutrients such as nitrogen and iron to survive. The iron and nitrogen found in fecal plumes is a byproduct of the diet consumed by whales. These nutrients are limiting factors, meaning they can limit the number of organisms in a population. Like all limiting factors, they directly affect the carrying capacity of an ecosystem and limit the population of phytoplankton. In the *Chlorella* Investigation, we tested our model to see if an increase in nutrients corresponded to an increase

in algae growth. Our data confirmed that nutrient-rich whale fecal plumes provide an important boost to the phytoplankton populations.

In the #WhaleLife task, we learned about the connections among organisms in a marine food web. The increased phytoplankton populations bolstered by the whale fecal plumes and kick-up of nutrients to the surface provide a plentiful food source for zooplankton. The phytoplankton (producers) make up the base of the food web and are eaten by the zooplankton (primary consumers). The zooplankton, in turn, are eaten by larger zooplankton and small fish (secondary consumers), larger fish (tertiary consumers), and so on up the food chain. In the Energy Flow in an Ecosystem task, we learned that energy is passed up the food web from the producers to the consumers. Because nearly 90% of energy is lost at each trophic level, the producers must be plentiful in order to support the energy needs of the food chain dependent on them. Phytoplankton are at the base of a complex aquatic food web, and the increase in their abundance provides energy to maintain larger populations and biomass of consumers at the higher trophic levels. As the populations of plankton and fish increase, more food becomes available for the whales. This, in turn, leads to the production of more whale fecal plumes. The release of these additional nutrients serves to amplify the nutrient source for the phytoplankton, creating a positive feedback loop in this marine ecosystem.

The story doesn't end here. We learned from the Trophic Cascades task that one organism can have far-reaching effects on an ecosystem. These effects are both direct and indirect, resulting in changes to populations of organisms and also having an impact on the abiotic, or nonliving, factors within the environment. It turns out that whales help indirectly sequester a significant amount of CO_2, a gas largely responsible for global climate change. We learned in the Carbon Cycle Game that carbon moves throughout the biosphere and geosphere through many processes. When whales come to the surface of the ocean, they release nutrient-rich fecal plumes, which stimulate phytoplankton growth. These phytoplankton not only feed other ocean organisms (krill and fish) but also are photosynthetic organisms that remove and store large quantities of atmospheric CO_2. Therefore, the greater the phytoplankton population, the more CO_2 is drawn out of the atmosphere. As CO_2 is a greenhouse gas that causes the atmosphere to heat up, the decline in the whale populations also led to less CO_2 being sequestered by phytoplankton that were nourished by whale plumes, which consequently contributed to global climate change.

Eliciting Ideas About the Phenomenon

The eliciting ideas about the phenomenon stage of MBI is about introducing the phenomenon and driving question, eliciting students' initial ideas about what might explain the phenomenon and answer the driving question, and constructing their initial models in small groups. We have provided a PowerPoint for this purpose that will lead you step-by-step through the process, which can be downloaded from the book's Extras page at *www.nsta.org/mbi-biology*. There are two products of this stage: an Initial Hypotheses List and student groups' initial models. Examples of each are shown in Figures 2.1 and 2.2 (p. 164).

Figure 2.1. Example Initial Hypotheses List

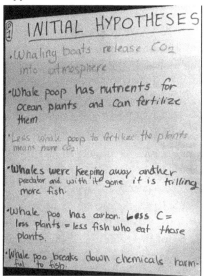

This stage of MBI is critically important, as it orients students to the MBI unit anchoring phenomenon using the driving question in a way that makes it compelling enough to spend several weeks across a unit working on explaining an event. Put more succinctly, this stage introduces the problem space in which the students will be working. Introducing the phenomenon should be done in a way that supports students in drawing on what they have learned previously in and outside of school as they begin to think about how to explain the driving question. Your priority should be to elicit students' ideas about the phenomenon, since it is important for them to continually think about the phenomenon and refine their everyday ways of thinking. Since eliciting students' ideas is a top priority, it's important to create a learning environment where students feel comfortable and are invited to offer ideas. This means you need to think carefully about how to get students to float (or put on the table for consideration) as many ideas as possible. At this stage, there are no right or wrong ideas. Everything (within reason!) is on the table.

Figure 2.2. Example Initial Model

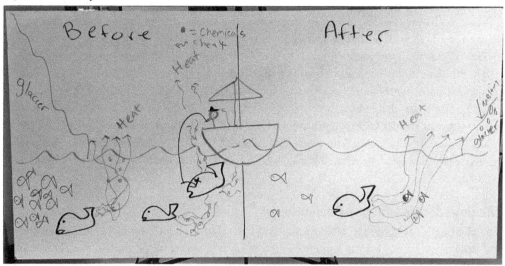

Negotiating Ideas and Evidence Through Tasks

The negotiating ideas and evidence through tasks stage of MBI makes up the majority of the unit. Each task is designed to introduce a key scientific idea or to reinforce an idea already raised by students in the class. Halfway through the unit, we recommend having groups revise their models and share out those revisions. PowerPoint templates are provided for the tasks as well as for the model revision; these can be downloaded from the book's Extras page at *www.nsta.org/mbi-biology*. Each task begins by introducing ideas for students to reason with while working on the task. During the task, students investigate or test the concepts using data. We recommend using the Back Pocket Questions provided in the Teacher Notes to press students' thinking while they are engaged in the task. Taken together, the tasks scaffold students' thinking as they co-construct a scientific explanation of the phenomenon. The general outline of this stage is as follows:

Task 10. #WhaleLife: Interdependence in Marine Communities

Task 11. Limiting Factors for Phytoplankton

Task 12. Energy Flow in an Ecosystem

Midunit Model Revision for Unit 2

Task 13. Trophic Cascades

Task 14. Carbon Cycle Game

Task 15. Testing Your Model: *Chlorella* Investigation

A Summary Table is used after each task to scaffold students' negotiation of ideas and how those ideas help explain the phenomenon. An example is provided in Table 2.2 (p. 166). Like the target explanation, this is included here to serve as a behind-the-scenes roadmap to help support you, the teacher, in responsively supporting students in negotiating similar responses by the end of the unit. It is not something to be shared with students but is provided for you to consider before each task to guide your instruction.

You can think of your role here as that of helping introduce ideas to students to support them in learning about and beginning to engage with the ideas through carefully designed sensemaking tasks. If important ideas are not introduced by the students, you can put them on the table or introduce them through the use of just-in-time direct instruction, short videos, or readings, among other strategies. Since the objective is to have students pick up and try out these ideas in carefully crafted tasks as they develop explanations, introducing your students to an idea early in the task is *not* giving them answers that will be confirmed in the task. Instead, you can think of this as introducing students to tools (i.e., disciplinary core ideas) that will be useful in a task in ways that will help them make sense of data, a simulation, an activity, or something in the world. In the end, it is important that you think of introducing new ideas in the early parts of this MBI stage as an opportunity for students to engage in sensemaking with new ideas to use, while you concurrently recognize the potential explanatory power of the new ideas. At the end of each task, the completion of a row of the Summary Table is an opportunity for students to think about what they have learned in the task and, together with peers, reason about how the newly introduced ideas can be applied to support their objective of explaining the unit anchoring phenomenon.

Table 2.2. Example Summary Table for Unit 2

Task	What we learned from this task	How it helps us explain the anchoring phenomenon
10. #WhaleLife: Interdependence in Marine Communities	Food webs are used to show the feeding relationships among organisms (producers, consumers, and decomposers) in a community. The web shows the flow of energy and matter from one trophic level to the next.	Phytoplankton provide chemical energy for small fish and krill, which provide chemical energy for whales in a marine food web. Phytoplankton gain energy to grow from sunlight.
11. Limiting Factors for Phytoplankton	Limiting factors are those factors that limit the growth of populations. Phytoplankton rely on sunlight, carbon dioxide, and nutrients (nitrogen, iron, and silica) in order to grow and reproduce.	The presence of nitrogen in the ocean is greater when whales are also present. The presence of nitrogen in fecal plumes increases the amount of phytoplankton in the ocean.
12. Energy Flow in An Ecosystem	The 10% rule shows us that the energy stored in biomass at each trophic level is lost/used as it flows to higher trophic levels. This means that populations at lower trophic levels have to be much larger to feed higher trophic levels.	A lot of energy/biomass must be produced by phytoplankton to support healthy fish populations.
13. Trophic Cascades	Trophic cascades result from removing or adding an organism to an ecosystem, leading to positive or negative direct or indirect effects that have far-reaching consequences in an ecosystem.	Whale hunting indirectly affects the number of fish in the ocean. The presence of fewer whales leads to a decrease in fecal plumes, decreasing the phytoplankton, decreasing krill and zooplankton, decreasing fish.
14. Carbon Cycle Game	Molecules of carbon cycle between the biotic and abiotic parts of the environment through carbon reservoirs via a number of different processes. Living things such as phytoplankton remove and store carbon from the atmosphere and thus have an impact on climate change.	Fewer whales mean less nitrogen available in the ocean, leading to less phytoplankton. Smaller populations of phytoplankton capture less CO_2 from the air.
15. Testing Your Model: *Chlorella* Investigation	Increasing the amount of Miracle-Gro leads to an increase in phytoplankton growth and reproduction.	Since whale plumes contain similar ingredients to Miracle-Gro, we now have evidence (data) that shows that an increase in whale populations (and consequently plumes) will have a positive effect on phytoplankton growth in the ocean.

Building Consensus

In the building consensus stage of an MBI unit, the whole class works to build consensus about the explanation of the phenomenon by finalizing the groups' models, comparing and contrasting those models as a whole class, and constructing a consensus checklist of the ideas and evidence (called the Gotta Have Checklist) that should be a part of students' final evidence-based explanations that make up the summative assessment of the unit. A PowerPoint is provided as a guide through this stage of the unit and can be downloaded from the book's Extras page at *www.nsta.org/mbi-biology*. Examples of a group's final model and the whole-class Gotta Have Checklist appear in Figures 2.3 and 2.4. In this unit, students will co-construct a Gotta Have Checklist before completing their final models.

Figure 2.3. Example Gotta Have Checklist

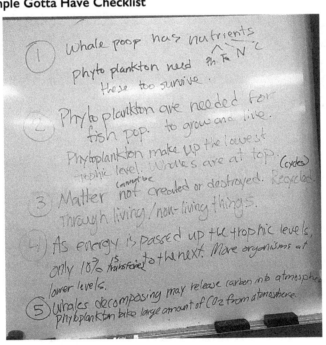

The checklist will vary slightly from class to class, but it may include the following:

- The ecosystem before whaling and the effects of whaling on the ecosystem
- Relevant organisms from the marine food web and the relationships among those organisms
- Relative numbers of organisms based on trophic pyramid
- Limiting factors that affect phytoplankton and the sources of these abiotic factors
- How phytoplankton remove carbon from the atmosphere
- How the accumulation of carbon dioxide in the atmosphere leads to climate change

Figure 2.4. Example Final Model

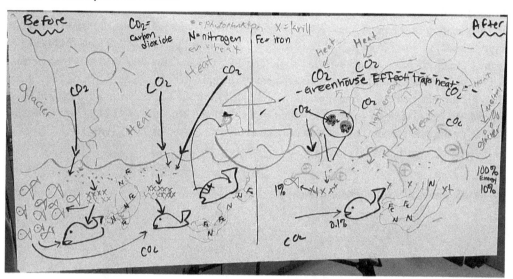

In Chapter 2, we described the scientific practice of modeling as a knowledge-building practice and models as the products of modeling. Students begin this stage by finalizing their models in groups of three or four, which they should think of as a continuation of the sensemaking that started in the groups' initial modeling experiences in the early days of the MBI unit and continued when they revisited these models midway through the unit. As with these other experiences, you can expect the practice of modeling to support groups in refining their explanations of the unit anchoring phenomenon. Here, negotiation and argumentation lead to refined ways of thinking about the phenomenon that may not have occurred had students not engaged in modeling this final time. The groups' models will offer insight into students' thinking that otherwise may not be accessible to you. At the same time, models as products of modeling also serve as artifacts that students can compare across groups in ways that are supported by your selection and sequencing of groups' sharing in whole-class sensemaking, so that differences can be foregrounded, negotiated, and resolved through argumentation, among other science practices.

Finally, the construction of final criteria or whole-class mapping of ideas included in the Gotta Have Checklist to the evidence collected from the various tasks completed in the second stage of MBI is a scaffold that will be important for individual students in writing evidence-based explanations. Importantly, the teacher-facilitated, co-constructed Gotta Have Checklist outlines what the class agrees should be included in students' explanations and is mapped to evidence identified across the unit tasks. This step is the last stage of building consensus. This is especially true because final group models will have been negotiated in both small and whole groups as a basis for the identification of the criteria that are needed for writing an evidence-based explanation.

Establishing Credibility

In MBI, students must argue for their ideas in writing. In the establishing credibility stage, they do this through written evidence-based explanations, peer review, and revision. A PowerPoint is provided as a guide through this stage of the unit and can be downloaded from the book's Extras page at *www.nsta.org/mbi-biology*. Writing scaffolds such as sentence stems may be necessary to jump-start some students' writing. Once the written explanations are complete, we recommend using the MBI Explanation Peer-Review Guide found in Appendix B. The target explanation on pages 162–163 is an example.

Writing evidence-based explanations is a sensemaking experience that students engage in early in this stage of MBI. This is another knowledge-building opportunity that, like engaging in modeling earlier in the unit, is a practice that leads to a product. And like models, which served as products that could provide you with insight into the way groups were thinking about explaining the anchoring phenomenon, the individually written evidence-based explanations afford you with an opportunity to assess individual students in terms of where they are in their attempts to explain the anchoring phenomenon. As students engage with others in this stage, this is an additional space for argumentation and negotiation as they recognize possible differences and details they may not have considered or may not yet agree with.

This practice of peer review needs to be supported, so we have provided the MBI Explanation Peer-Review Guide in Appendix B, which students can use to provide feedback to each other. The peer-review guide contains questions designed to focus student attention on the most important aspects of explanations and to encourage students not only to discuss what they figured out and how they know what they know but also to best communicate what they have learned to others in a way that is clear, complete, and persuasive. During the peer review, you should consider and make explicit how explanations, like models, are refined over time through negotiation within a community. The MBI unit culminates at the end of this stage as students use the feedback they have received to make final revisions that represent how they are thinking about the unit anchoring phenomenon. The final evidence-based explanations serve as one measure of what students have learned about engaging in science practices to use disciplinary core ideas and crosscutting concepts to explain the phenomenon.

Hints for Implementing the Unit

- The issue of whaling may be controversial or emotional for some students. This may lead to discussions that could sidetrack the initial phase of the unit (eliciting ideas about the phenomenon). Let students know that the class will not be discussing the moral or ethical questions regarding whaling. Instead, they will focus on the effects of whaling on marine ecosystems.

- We recommend providing opportunities to connect students' personal experiences and community resources with the phenomenon and topic of this unit. While students may not have had any personal experiences with the phenomenon, we suggest holding discussions throughout the unit to elicit their personal connections. Think about possible connections in your community that you can bring to the conversation as well. These kinds of connections will make the unit personally meaningful to your students, increasing their motivation to engage in the tasks. In our experience, much of this happens during the eliciting ideas about the phenomenon stage of the unit. However, you should prompt students to connect to the topic during all phases of the unit.

- There may be times during the unit when students make connections to injustices or disparities in their communities or raise emotional responses to the topics. We suggest preparing to address young people's questions and desire for activism so science practices are not portrayed as disconnected from the social and cultural contexts of students' real-world experiences.

Targeted NGSS Performance Expectations

- **HS-LS2-1.** Use mathematical and/or computational representations to support explanations of factors that affect carrying capacity of ecosystems at different scales.

- **HS-LS2-2.** Use mathematical representations to support and revise explanations based on evidence about factors affecting biodiversity and populations in ecosystems of different scales.

- **HS-LS2-3.** Construct and revise an explanation based on evidence for the cycling of matter and flow of energy in aerobic and anaerobic conditions.

- **HS-LS2-4.** Use mathematical representations to support claims for the cycling of matter and flow of energy among organisms in an ecosystem.

- **HS-LS2-5.** Develop a model to illustrate the role of photosynthesis and cellular respiration in the cycling of carbon among the biosphere, atmosphere, hydrosphere, and geosphere.

- **HS-LS2-6.** Evaluate the claims, evidence, and reasoning that the complex interactions in ecosystems maintain relatively consistent numbers and types of organisms in stable conditions, but changing conditions may result in a new ecosystem.

- **HS-LS2-7.** Design, evaluate, and refine a solution for reducing the impacts of human activities on the environment and biodiversity.

- **HS-LS2-8.** Evaluate the evidence for the role of group behavior on individual and species' chances to survive and reproduce.

Eliciting Ideas About the Phenomenon Stage Summary

The first stage of MBI, eliciting ideas about the phenomenon (Figure 2.5), involves introducing the anchoring phenomenon and driving question, eliciting students' initial ideas and experiences that may help them develop initial explanations of the phenomenon, and developing initial models of the phenomenon based on those current ideas. Before beginning this stage, review the relevant sections in Chapters 1 and 2 for specifics. See Figure 2.1 on page 164 for an example of an Initial Hypotheses List for this unit. We have provided a PowerPoint template to assist with this stage, which can be downloaded from the book's Extras page at *www.nsta.org/mbi-biology*.

Figure 2.5. Eliciting Ideas About the Phenomenon Stage Summary

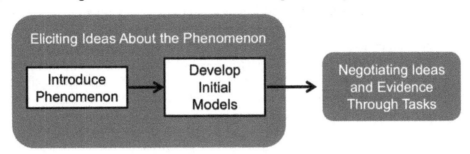

Introduce the Phenomenon

We begin this stage by introducing the phenomenon in an engaging way, such as with stories, videos, demonstrations, or even short activities. The goal is to provide just enough information for students to begin to reason about the phenomenon, without providing too much of the explanation. While we are introducing the phenomenon, we ask questions to keep students engaged and make sure they are paying attention to the important aspects that can begin to help them explain the phenomenon. The introduction ends with the driving question of the unit, which we have found helps focus their thinking on the development of a causal explanation for the phenomenon, or a "why" answer.

It is important to spend time eliciting not only students' scientific ideas about the phenomenon but also their personal connections to the phenomenon. What personal experiences do they have that help them connect to the phenomenon? Even though the phenomenon of the unit may not be directly connected to your community, does the community have similar phenomena or specific resources that you can bring into the discussion? The more connected students feel to the unit phenomenon, the more engaged and motivated they will be.

We then get students into their groups and facilitate the first discussion to try to answer the driving question with just the resources they brought with them—the ideas, experiences, and cultural resources they have gained both in and outside of the classroom. These ideas may be fully formed, partially correct, or fully incorrect in terms of our canonical knowledge of science. However, we make it clear that all ideas are considered equally valid at this point in time, as we realize these are the ideas students put into play when they think about the phenomenon we have introduced.

Once student groups have discussed their ideas, we facilitate a class discussion to compare and contrast the ideas generated by each group. As ideas are presented, they are put on our first public record, which we call the Initial Hypotheses List. We use this list throughout the unit to keep track of the changes in students' thinking as they work toward a final evidence-based explanation for the anchoring phenomenon. We consider all ideas to be valid at this stage, before students use other resources to begin making sense of how or why something happens. The Initial Hypotheses List is also useful for the next task in this stage, initial model construction, especially since it offers students additional ideas beyond those they initially had either individually or in their small groups.

Develop Initial Models

If students are not experienced with modeling, it is worth providing a brief introduction and example. We have included an example in the PowerPoint for this stage, which can be downloaded from the book's Extras page at *www.nsta.org/mbi-biology*. Once the class is ready to begin modeling, we give each group a sheet of 11 × 17 inch paper and ask the groups to each make a model of their initial hypothesis. Sometimes they choose their own original hypotheses, while other times they are influenced by their peers' ideas and adopt one of them instead. As the groups work on constructing their models, you should walk around asking clarifying questions and pushing them to be as specific as possible. Once the models are ready, it is important to have students share ideas across them. There are a number of ways you might run these share-out sessions. We often collect and present the models on a document camera at the end of the first day. Groups can provide one- or two-sentence summaries of the initial hypotheses they have represented in their models. We point out interesting ideas and ways in which they have represented these ideas. For example, we may call attention to the fact that a group labeled the arrows, which made the model more understandable, and that another group used a zoom-in window to show what was happening at a different scale. At the end of the first day and this first stage, we have elicited ideas across the class, and the groups' initial hypotheses and models will act as a starting point for the rest of the unit.

Negotiating Ideas and Evidence Through Tasks Stage Summary

The next stage of MBI, negotiating ideas and evidence through tasks (Figure 2.6), takes up the majority of the unit. The tasks are designed to introduce or extend important science ideas students need as they construct their evidence-based explanations of the phenomenon. Before beginning the tasks, we suggest that you review the relevant sections in Chapters 1 and 2 for specifics. Each task consists of the following:

- A **Teacher Notes** section that provides an overview of the activity and science content important to the task. Here, we provide guidance on conducting the task with details on the procedure, a list of required materials and preparation, any necessary safety precautions, suggested Back Pocket Questions, and an example Summary Table entry. At the end of this section are further hints for implementing the task and possible extensions, if desired.

- A **Student Handout** to be given to groups as they engage in the task. The handout provides an introduction to the task and content, an initial ideas section to frame the learning before they begin, and detailed instructions and work space. It ends with a section called What We Figured Out, which is designed to scaffold student responses to include in the Summary Table as a whole class. We suggest having students fill out this section after completing the task and the post-task discussion.

- A **PowerPoint** Negotiating Ideas and Evidence Through Tasks template that you can adapt for each task, which can be downloaded from the book's Extras page at *www.nsta.org/ mbi-biology*.

- A Midunit Model Revision reminder roughly halfway through the unit.

Figure 2.6. Negotiating Ideas and Evidence Through Tasks Stage Summary

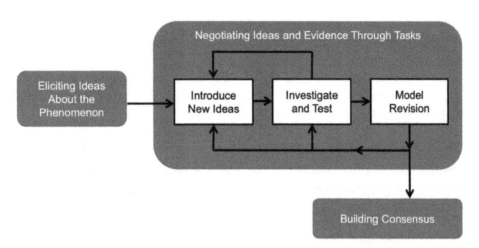

Teacher Notes

Task 10. #WhaleLife: Interdependence in Marine Communities

Purpose

During the eliciting ideas about the phenomenon phase, students modeled their initial ideas about how whale plumes have affected fish populations and climate change. In this task, they construct a marine food web to show the feeding relationships among whales and the other organisms in their community. In the next task, students will examine factors that may limit the growth of the phytoplankton population in this community.

Important Life Science Content

Most students have been introduced to food webs in middle school. Food webs illustrate the flow of matter and energy in a community. They are composed of three trophic levels: producers, consumers, and decomposers. Producers form the lowest trophic level of a food web. Most students will be able to identify plants as a producer but may never have heard of phytoplankton. In a marine food web, the main producers are phytoplankton, which can include microscopic marine algae and bacteria that use the sun as a source of energy to make food. When primary consumers eat these producers, chemical energy is transferred. Consumers can also eat other consumers (secondary consumers, tertiary consumers, and so on). Decomposers recycle matter from dead producers and consumers back to the water or soil.

Scientific Ideas That Are Important to Think About During This Task

- Food webs show the connections among organisms in a community and reveal the movement of energy and matter.
- The trophic level of an organism is the position it occupies in a food web, or the number of steps it is from the start of the chain. A food web starts with primary producers such as plants at trophic level 1, can include herbivores at level 2 and carnivores at level 3 or higher, and typically finishes with apex predators at level 4 or 5.
- Food webs are one way to reveal the interdependence of organisms within ecosystems.

Timeline

Approximately one class period.

Materials and Preparation

The items needed for this investigation are listed in Table 2.3. Print a set of Food Web Cards for each group from *https://d43fweuh3sg51.cloudfront.net/media/media_files/marine_science_Food_Web_Cards.pdf*.

TASK 10. #WHALELIFE: INTERDEPENDENCE IN MARINE COMMUNITIES

Table 2.3. Required Materials for Task 10

Item	Quantity
Set of Food Web Cards	1 per group
Poster paper	1 per group
Highlighters in 4 different colors	1 set per group
Student Handout	1 per student

Safety Precautions

This task does not require any specific safety precautions.

Procedure

This lesson plan is only a suggestion. It is included here to illustrate how you can facilitate student thinking during this task. We encourage you to modify this lesson plan by asking different questions, using different examples, and providing different scaffolds as appropriate to better meet the needs of students in your class.

Introduction of the Task (10 minutes)

1. Introduce the concept of food chains and food webs to students. Key terms such as *producer*, *consumer*, and *decomposer* should be discussed. Use a terrestrial ecosystem as an example, as students may be more familiar with the organisms. This terrestrial food web can serve as a reference for students to look to while doing the task.

2. Be sure to discuss the arrows used in food chains and food webs. The arrows are used to illustrate how energy and matter move from one organism to the next: producer → primary consumer → secondary consumer → tertiary consumer. Students often want to draw the arrows in the opposite direction, indicating which organism is eating the other organisms. This should be addressed before beginning the task.

3. To help students build background knowledge, show students pictures of phytoplankton and krill and explain what types of organisms they are.

4. Review with students the basic material and intellectual procedures of the task.

Analyzing and Interpreting Data (30 minutes)

1. Monitor students as they work with the data to construct food chains. Provide feedback as needed.

2. As students work, move from group to group and check in with them. It's important to ask them questions that will help them connect what they are doing to the goal of the

task and the anchoring phenomenon. (See the Back Pocket Questions for Task 10 below for some suggestions).

Putting Ideas on the Table (10 minutes)

1. Assess students' understanding of food webs using discussion points on the Task Wrap-Up slide in the PowerPoint we have provided, which can be downloaded from the book's Extras page at *www.nsta.org/mbi-biology*.

2. Introduce the crosscutting concept of energy and matter, again using the Task Wrap-Up slide.

Adding Information to the Summary Table (10 minutes)

1. Give students 5 minutes to decide what to add to the Summary Table at the end of the Student Handout.

2. Have one student from each group share what the group figured out, how they know (their evidence for what they figured out), and how this information will help them explain the anchoring phenomenon.

3. Once each group has shared, ask the entire class to decide what should be added to each column of the class Summary Table for Task 10. Help students reach consensus about what to add to the Summary Table. Only add an idea to the Summary Table if everyone in the class agrees with that idea.

Back Pocket Questions

As students work in groups, it is important to engage with each group to help press and extend students' thinking around the ideas at play in this task. Following are some example questions you might ask:

1. Helping students get started: What do you notice about the relationship of the whales to the other organisms in their community? What questions do you have about marine communities? Food webs?

2. Pressing further: What do you think would happen if [organism X] were removed from the community? How do you think this relates to how whales affect the food web?

3. Following up: What makes you think that? Can you say more?

Filling Out the Summary Table

Table 2.4 includes examples of the responses students may come up with when they fill out the Summary Table. This is provided here only as a behind-the-scenes roadmap and is not meant to be shared with students.

TASK 10. #WHALELIFE: INTERDEPENDENCE IN MARINE COMMUNITIES

Table 2.4. Example Summary Table for Task 10

What we learned	How it helps us explain the phenomenon
Food webs are used to show the feeding relationships among organisms (producers, consumers, and decomposers) in a community.	Phytoplankton provide chemical energy for small fish and krill, which provide chemical energy for whales in a marine food web.
The web shows the flow of energy and matter from one trophic level to the next.	Phytoplankton gain energy to grow from sunlight.

Hints for Implementing This Task

- Use the highlighters to help students visualize the trophic levels in a food web. This concept will be explored further in subsequent activities.

Possible Extensions

- If your students need more support, have them complete a reading about food chains and food webs. You could also show students a nature video that includes information about the ecosystem in which whales live.

- This task was adapted from the PBS Learning Media task "Modeling Marine Food Webs and Human Impacts on Marine Ecosystems," which provides additional materials at *https:// az.pbslearningmedia.org/resource/marinesci-sci-foodwebs/food-webs*.

- If students are familiar with this food web, have them investigate an alternative marine ecosystem, such as hydrothermal vents.

Student Handout

TASK 10. #WHALELIFE: INTERDEPENDENCE IN MARINE COMMUNITIES

Introduction

Food webs illustrate the flow of matter and energy in a community. They are composed of three groups of organisms: producers, consumers, and decomposers. *Producers* form the lowest level of a food web and consist of organisms that can produce their own food. In a marine food web, the main producers are phytoplankton. *Phytoplankton* are a community of marine organisms, mostly microscopic marine algae and bacteria, that use the Sun as a source of energy to make food. When *primary consumers* eat these producers, chemical energy and matter are transferred. Consumers can also eat other consumers (secondary consumers, tertiary consumers, quaternary consumers, and so on). *Decomposers* recycle matter from dead organisms back to the water or soil. As a way of organizing food webs, you can think in terms of *trophic levels*, the positions organisms occupy in a food web. Producers make up the first trophic level, followed by the different levels of consumers. During this task, be sure to pay attention to each of these types of organisms.

Initial Ideas

Before beginning this task, take a few minutes to think about food webs and where whales fit in. In what ways do you think whales might be dependent on other organisms in the ocean?

How are food webs used to show the flow of energy and matter among organisms of a community?

TASK 10. #WHALELIFE: INTERDEPENDENCE IN MARINE COMMUNITIES

Your Task

Your task is to begin exploring interdependence in marine communities by constructing a marine food web to show the feeding relationships among whales, fish, and other organisms in a whale's community. In the next task, you will build on what you learn by examining factors that might limit the growth of the phytoplankton population in this community.

1. Fill in the missing boxes in the Trophic Levels table below, using information from the introduction and your cards to help.

Trophic Levels

Trophic level	Role	Examples (from cards)
	uses sunlight to make food	
Primary consumer	herbivore: eats only producers	
Secondary consumers	carnivore: eats animals *or* omnivore: eats animals and producers	
Tertiary consumers	carnivore: eats animals *or* omnivore: eats animals and producers	
	recycles dead matter into nutrients	bacteria

2. Make a food chain consisting of 3, 4, or 5 cards (*producers and consumers*) on a large piece of paper. *One of the chains should include the whale!*

3. Place the producer at the bottom and the tertiary (or quaternary) consumer at the top of each chain.

4. Connect the cards with arrows to show where the energy travels in the food chain (*from the organism that is being eaten TO the organism that is eating it*).

5. Repeat steps 2, 3, and 4 until you have built four different food chains. (Make sure at least one chain contains a whale and another contains a fish.)

6. Once you have sorted out the cards and made the connections, record your food chains in the Marine Food Chains table on the following page.

Marine Food Chains

Food chain #1	Food chain #2	Food chain #3	Food chain #4
(Consumer)	(Consumer)	(Consumer)	(Consumer)
(Producer)	(Producer)	(Producer)	(Producer)

National Science Teaching Association

TASK 10. #WHALELIFE: INTERDEPENDENCE IN MARINE COMMUNITIES

7. Using a different-colored highlighter for each level, label the producer, primary consumer, secondary consumer, tertiary consumer, quaternary consumer (*if present*), and decomposer in each food chain in the table. Use the underlined words as your key.

8. Ecosystems like the one we are examining here are far more complex than the food chains you just created. In reality, these chains are connected in large food webs that give a more realistic picture of how energy and matter move through ecosystems. In the space below, create and record a food WEB using the names of organisms from the cards. Your food web should include color-coded organisms with the correct highlighter color.

MY MARINE FOOD WEB

9. What do you observe about the role of whales and fish in your food web?

10. What do you predict the effects would be if you removed the whales from your food web? Give an example of an effect.

Challenge

- Add to your food web. Add some of the other organisms from the cards into your food web. Highlight them to show their trophic levels.
- Place the decomposers in your food web. Describe how their location is different from the other organisms.

TASK 10. #WHALELIFE: INTERDEPENDENCE IN MARINE COMMUNITIES

Some Useful Ideas From My Teacher

You can keep track of useful ideas from your teacher in the space below.

What We Figured Out

Now that you have completed this task, take a few minutes to fill out the Summary Table below with the other students in your group. This table will help you keep track of what you figured out during the task. You will then have an opportunity to share your ideas as we fill out our class Summary Table.

Summary Table

What we learned	How we know

How it helps us explain the phenomenon

Teacher Notes

Task 11. Limiting Factors for Phytoplankton

Purpose

In the previous task, students learned that phytoplankton are the base of the marine food chain. The goal of this two-part task is to have students investigate the various living and nonliving factors that limit phytoplankton growth and therefore affect the rest of the marine food web. In Part 1, students research four abiotic limiting factors—sunlight, carbon dioxide, nitrogen, and iron—to figure out why they are important to phytoplankton and where they come from. In Part 2, students look at the relationship between whales and phytoplankton, the availability of some of these limiting factors, and their effects on phytoplankton.

Important Life Science Content

Ecosystems include communities of organisms and the abiotic (nonliving) components of their environment. For an ecosystem to maintain itself, there must be enough available resources, both biotic and abiotic. If there is a change in the amount of these resources, the populations within the ecosystem will be affected.

Phytoplankton are a community of microscopic marine organisms that carry out photosynthesis to make food. Photosynthesis requires sunlight as an energy source to convert carbon dioxide and water into glucose and oxygen. In the photic zone of the ocean, sunlight is abundant from the penetration of the Sun's rays into the upper 260 feet (80 m) of the ocean. Carbon dioxide is absorbed from the atmosphere into the ocean, providing phytoplankton with the carbon necessary for photosynthesis to take place. However, the raw materials for photosynthesis alone are not enough to sustain phytoplankton growth. Additionally, populations of phytoplankton require nutrients such as nitrogen and phosphorus, trace elements such as iron, and compounds such as silica to allow for carbon fixation and to help them grow. The availability of these inorganic elements limits the size of phytoplankton populations. The largest amount of phytoplankton the ecosystem can support is referred to as the ecosystem's carrying capacity.

Scientific Ideas That Are Important to Think About During This Task

- Biotic (living) and abiotic (nonliving) factors enable and limit population growth of organisms within ecosystems.
- Specific factors that limit the growth of populations are referred to as limiting factors.

Timeline

Approximately one class period.

Materials and Preparation

The items needed for this investigation are listed in Table 2.5.

Table 2.5. Required Materials for Task 11

Item	Quantity
Computer or tablet with internet connection	1 per student
Student Handout	1 per student

Safety Precautions

This task does not require any specific safety precautions.

Procedure

This lesson plan is only a suggestion. It is included here to illustrate how you can facilitate student thinking during this task. We encourage you to modify this lesson plan by asking different questions, using different examples, and providing different scaffolds as appropriate to better meet the needs of students in your class.

Part 1 (30 minutes)

1. Introduce limiting factors and the task using the PowerPoint template we have provided, which can be downloaded from the book's Extras page at *www.nsta.org/mbi-biology*.

2. Have students work collaboratively in groups to answer the Initial Ideas questions at the top of the Student Handout.

3. The first part of this activity consists of internet research on four important limiting factors: sunlight, carbon dioxide, nitrogen, and iron. Relevant online sources are listed at the end of the handout. Depending on group size, this could be organized in one of two ways:

   a. If groups consist of four students, conduct a jigsaw of the information within the group by assigning a limiting factor to each student. Students then share out their findings within their groups. Have students fill in the table on the Student Handout.

   b. If groups have more or fewer than four students, conduct a jigsaw within the whole class. Assign the four limiting factors equally across the class. Provide time for each limiting factor expert group to discuss what it found. Then break students into jigsaw groups, ensuring that at least one student represents each limiting factor within each group. Allow enough time for all students to present their segments to the groups and to fill in the table on the Student Handout. Then bring the original groups back together.

Part 2 (30 minutes)

1. In the second part of the task, students look at real data from a study of the impact of whales on nitrogen levels in the Gulf of Maine.

2. Provide time for group members to analyze the graphs and answer the questions beside each graph as well as the wrap-up questions after the graphs.

3. If needed, model with students how to annotate a graph. Project the first graph on the board and annotate it as a group, then let students do the same individually or with partners for the second graph.

Adding Information to the Summary Table (10 minutes)

1. Give students 5 minutes to decide what to add to the Summary Table at the end of the Student Handout.

2. Have one student from each group share what the group figured out, how they know (their evidence for what they figured out), and how this information will help them explain the anchoring phenomenon.

3. Once each group has shared, ask the entire class to decide what should be added to each column of the class Summary Table for Task 11. Help students reach consensus about what to add to the Summary Table. Only add an idea to the Summary Table if everyone in the class agrees with that idea.

Back Pocket Questions

As students work in groups, it is important to engage with each group to help press and extend students' thinking around the ideas at play in this task. Following are some example questions you might ask:

1. Helping students get started: What do you notice about the different limiting factors?

2. Pressing further: How do you think these limiting factors for phytoplankton affect the whales and fish in the marine ecosystem?

3. Following up: What makes you think that? Can you say more?

Filling Out the Summary Table

Table 2.6 includes examples of the responses students may come up with when they fill out the Summary Table. This is provided here only as a behind-the-scenes roadmap and is not meant to be shared with students.

Table 2.6. Example Summary Table for Task 11

What we learned	How it helps us explain the phenomenon
Limiting factors are those factors that limit the growth of populations. Phytoplankton rely on sunlight, carbon dioxide, and nutrients (nitrogen, iron, and silica) in order to grow and reproduce.	The presence of nitrogen in the ocean is greater when whales are also present. The presence of nitrogen in fecal plumes increases the amount of phytoplankton in the ocean.

Possible Extensions

- The article "Scientists: Whale Poop Is Vital to Ocean's Carbon Cycle; 'Huge Amounts of Iron'" at *www.underwatertimes.com/news.php?article_id=52937108061* provides some data about the effect of whales on iron availability and recycling in the southern ocean. It is a good supporting resource to further discuss the role of whale plumes in providing iron.

- The case study "Is Iron Fertilization Good for the Sea?" at *http://sciencecases.lib.buffalo.edu/collection/detail.html?case_id=560&id=560* (access to this resource requires a paid subscription) describes the role of iron in the productivity of phytoplankton and its impacts on other trophic levels in the ocean.

- You may want to give the idea of carrying capacity more emphasis within this task or through an additional task.

Student Handout

TASK 11. LIMITING FACTORS FOR PHYTOPLANKTON

Introduction

In the previous task, you learned that phytoplankton are the main producers of the marine food web. All organisms in a marine community depend on phytoplankton either directly or indirectly for energy. There are factors that might limit the growth and reproduction of phytoplankton populations, and one is related to whale plumes.

Initial Ideas

Before beginning this task, take a few minutes to think about the limiting factors for phytoplankton. What do phytoplankton need to survive?

What is in whale fecal plumes that may affect phytoplankton?

Your Task

Your task is to examine factors that might limit the growth and reproduction of phytoplankton populations. Then you will analyze data from a real scientific study to figure out what is in whale plumes that affects phytoplankton. In the next task, you will design an investigation to test your ideas.

Part I

Today, the class will do a jigsaw activity to investigate relevant limiting factors to our whale phenomenon. Your teacher will explain how the jigsaw will be done. Be sure to record information on each factor in the Limiting Factors table below. See the Online Resources section at the end of this handout for useful websites that will help you conduct your research.

Limiting Factors

Limiting factor	Why is it important for phytoplankton?	Where does it come from? Where is it found in the ocean?
EXAMPLE: Silica	Many species of phytoplankton have hard outer shells (actually, their cell walls). Silica is important in making these shells. Without silica, phytoplankton won't have all the materials they need to grow.	Silica is contained in rocks, dust, and mud that blows into the ocean from land or from mud in rivers. Can be found in the top, middle, and bottom of the ocean.
Sunlight		
Carbon dioxide (CO_2)		
Nitrogen (N)		
Iron (Fe)		

Part 2

Examine the data in Figures SH11.1 and SH11.2 to look for patterns that will help you figure out how the overharvesting of whales led to a collapse in fish populations and contributed to an increase in carbon in the atmosphere. These data come from a scientific study of the impact of whales on nitrogen levels in the Gulf of Maine.

What patterns do you notice in the graphs in Figures SH11.1a and b? Make at least *three annotations* on each graph.

What do the graphs tell you about the relationship between nitrogen (N) flux (movement) and whales, presently and before commercial hunting?

Figure SH11.1. The Flux of Nitrogen in the Gulf of Maine (a) at Present and (b) Before Commercial Hunting

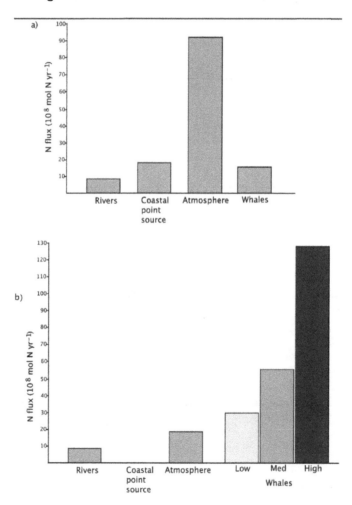

Note: N flux = movement (flow) of nitrogen.

National Science Teaching Association

TASK 11. LIMITING FACTORS FOR PHYTOPLANKTON

What patterns do you notice in the graph in Figure SH11.2? Make at least *three annotations.*

What does the graph tell you about the relationship between whale plumes and nitrogen availability?

Figure SH11.2. Shipboard Incubation Time-Course Experiments on Humpback Whale Samples Collected on Stellwagen Bank, Gulf of Maine

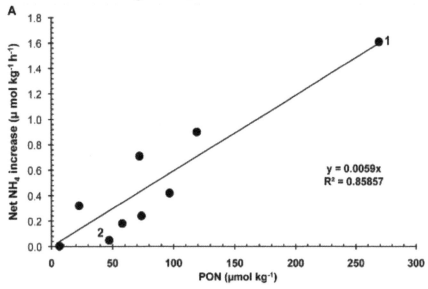

Note: PON = particles of nitrogen, a measure of how much whale fecal matter is in the water sample from the ocean; Net NH_4 = amount of ammonium, a form of nitrogen that phytoplankton can use.

How is nitrogen (N) a limiting factor for phytoplankton?

In this study, what impact did whales have on nitrogen levels in the Gulf of Maine?

Some Useful Ideas From My Teacher

You can keep track of useful ideas from your teacher in the space below.

What We Figured Out

Now that you have completed this task, take a few minutes to fill out the Summary Table below with the other students in your group. This table will help you keep track of what you figured out during the task. You will then have an opportunity to share your ideas as we fill out our class Summary Table.

Summary Table

What we learned	How we know

How it helps us explain the phenomenon

Online Resources for Part 1

Sunlight

- *https://oceanservice.noaa.gov/facts/light_travel.html*
- *https://nhpbs.org/natureworks/nwepphotosynthesis.htm*
- *www.extremescience.com/sunlight-zone.htm*

Carbon dioxide

- *www.scienceforthepublic.org/earth/how-plankton-blooms-absorb-co2*
- *https://climatekids.nasa.gov/ocean* (just read "How does the ocean soak up CO_2?")
- *https://whatsyourimpact.org/greenhouse-gases/carbon-dioxide-emissions*

Nitrogen

- *http://klimat.czn.uj.edu.pl/enid/2__Oceanic_nutrients/-_Phytoplankton_and_nutrients_1vf.html*
- *www.khanacademy.org/science/biology/ecology/biogeochemical-cycles/a/the-nitrogen-cycle*

Iron

- *http://klimat.czn.uj.edu.pl/enid/2__Oceanic_nutrients/-_Iron_in_the_oceans_1vv.html*
- *www.gardeningknowhow.com/garden-how-to/soil-fertilizers/iron-for-plants.htm*
- *www.whoi.edu/press-room/news-release/dissolved-iron*
- *www.whoi.edu/wp-content/uploads/2015/02/OceanicIronSources_750_372757.jpg*

Teacher Notes

Task 12. Energy Flow in an Ecosystem

Purpose

In the previous task, students learned about the food chains that make up an oceanic food web. In this task, they examine the flow of energy through these food chains to determine the impact of the 10% rule on the amount of energy available at each level of the food chain. In the next task, students will further investigate the relationships between organisms and their ecosystems as they study the effect of trophic cascades.

Important Life Science Content

When energy enters a trophic level, some of it is stored as biomass, as part of organisms' bodies. This is the energy that's available to the next trophic level, since only energy stored as biomass can get eaten. In food webs, there is a loss of energy from one trophic level to the next. On average, only about 10% of the energy that is stored as biomass at any trophic level is transferred to the next level. The remaining energy is lost through metabolic processes as heat. This is known as the 10% rule and helps determine the number of organisms that can be sustained at each trophic level of an ecosystem. Since 90% of energy is lost at each trophic level, this means that lower trophic levels must have much larger populations to support smaller populations above. This is seen in the energy pyramid in Figure 2.7.

Figure 2.7. Energy Pyramid

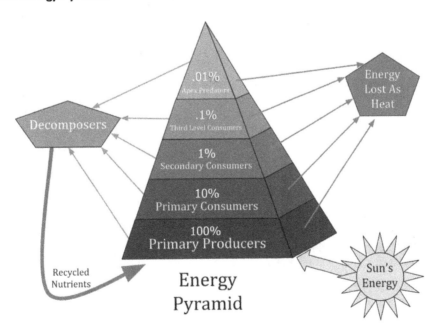

Scientific Ideas That Are Important to Think About During This Task

- Biomass is the total mass of organisms at a trophic level and thus represents the amount of stored energy available to be transferred up to the next trophic level.
- On average, most of the energy stored in biomass at each trophic level is released as heat or used by decomposers, leaving only about 10% of energy to be transferred up to the next trophic level. This is known as the 10% rule.
- The 10% rule explains why there are usually far more producers than primary consumers, more primary consumers than secondary consumers, and so on.

Timeline

Approximately one class period.

Materials and Preparation

The items needed for this investigation are listed in Table 2.7. Topographic maps of public land local to or near your area (e.g., a national park or BLM land) can be downloaded from many websites, such as *https://ngmdb.usgs.gov/topoview/viewer* (see Figure 2.8 for an example). We recommend using a 1:24,000 scale. Each group should receive a different topo map from the surrounding area to work with. Make calculators available to students as needed.

Table 2.7. Required Materials for Task 12

Item	Quantity
Topographic map	1 per group
Calculator	1 per student
Student Handout	1 per student

Safety Precautions

This task does not require any specific safety precautions.

Procedure

This lesson plan is only a suggestion. It is included here to illustrate how you can facilitate student thinking during this task. We encourage you to modify this lesson plan by asking different questions, using different examples, and providing different scaffolds as appropriate to better meet the needs of students in your class.

Figure 2.8. Topographic Map Example

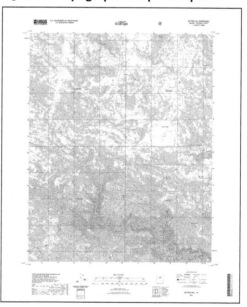

Putting Ideas on the Table (50 minutes)

1. Provide student groups with topographic maps and information on how many deer and mountain lions can live in their area. Have students calculate the acres of usable land on their map, then represent this graphically.

2. Give students conversions for food into calories. Ask them to try to graphically represent the 90–99% energy loss at each level of the food chain. Give them no further guidance on the graph or mode of representation.

3. Sequence share-outs. The objective is to demonstrate the difference between the pyramid of numbers and the pyramid of energy.

Adding Information to the Summary Table (10 minutes)

1. Give students 5 minutes to decide what to add to the Summary Table at the end of the Student Handout.

2. Have one student from each group share what the group figured out, how they know (their evidence for what they figured out), and how this information will help them explain the anchoring phenomenon.

3. Once each group has shared, ask the entire class to decide what should be added to each column of the class Summary Table for Task 12. Help students reach consensus about what to add to the Summary Table. Only add an idea to the Summary Table if everyone in the class agrees with that idea.

Back Pocket Questions

As students work in groups, it is important to engage with each group to help press and extend students' thinking around the ideas at play in this task. Following are some example questions you might ask:

1. Helping students get started: What patterns are you noticing on your topographic maps?

2. Pressing further: How does the idea of energy relate to this map? Do your findings support your hypothesis? Why or why not?

3. Following up: What makes you think that? Can you say more?

Filling Out the Summary Table

Table 2.8 (p. 198) includes examples of the responses students may come up with when they fill out the Summary Table. This is provided here only as a behind-the-scenes roadmap and is not meant to be shared with students.

Table 2.8. Example Summary Table for Task 12

What we learned	How it helps us explain the phenomenon
The 10% rule shows us that the energy stored in biomass at each trophic level is lost/used as it flows to higher trophic levels. This means that populations at lower trophic levels have to be much larger to feed higher trophic levels.	A lot of energy/biomass must be produced by phytoplankton in order to support healthy fish populations.

Hints for Implementing This Task

- We recommend that students calculate deer and mountain lions using decimals (e.g., 12.5 mountain lions).

- Calculating usable area on the topo maps can be challenging. Students may need to simply estimate the usable area.

Possible Extensions

- The 10% rule can be demonstrated during the introduction to this task. Have four volunteer students come up to the front of the class, and provide each with a cup. You will also need a graduated cylinder with 1 liter of water and an eyedropper. Explain that you are the Sun and will distribute energy to the students, who are a producer, a secondary consumer, a tertiary consumer, and a decomposer. Start by giving 10% (10 ml) to the producer. Then the producer gives 10% (1 ml) to the secondary consumer. The secondary consumer gives 10% (0.1 ml) to the tertiary consumer. Finally, the tertiary consumer gives 10% (0.01 ml) to the decomposer.

Student Handout

TASK 12. ENERGY FLOW IN AN ECOSYSTEM

Introduction

In food chains, there is a loss of energy from one trophic level to the next. As little as 10% of the energy at any trophic level is transferred to the next level. The remaining percentage is lost through metabolic processes as heat. This is known as the 10% rule and determines the number of organisms that can be sustained at each trophic level of an ecosystem.

Initial Ideas

Before beginning this task, take a few minutes to think about the flow of energy in food chains. Where do you find energy in a trophic level?

On the African plains, why are there far more gazelles than cheetahs, their predators?

Thinking about the 10% rule in an ecosystem, what should you expect to see in terms of numbers of predator and prey? Why?

Your Task

Your task is to examine the flow of energy through food chains with phytoplankton at their base. To accomplish this task, you will use topographic maps and the assumptions below to *calculate* the number of deer and mountain lions that can live in your area. Remember to consider *only* the habitat that deer can live on (such as open land). Deer cannot survive well in urban development. Use your data to *test* the 10% rule by creating a graph. Your group can decide how you want to represent the data you found, but be prepared to share your reasoning with the class.

Assumptions

- One square mile of deer habitat produces 760 pounds of grass each year.
- One deer eats approximately 3,650 pounds of grass each year.
- The average deer weighs 150 pounds.
- One mountain lion eats approximately 1,125 pounds of deer meat each year. However, the mountain lion will eat only 50% of each deer it kills.
- The average mountain lion weighs 145 pounds.
- Calories in food:
    - 1 pound of grass = 450 calories (kcal)
    - 1 pound of deer meat = 700 kcal
    - 1 pound of mountain lion meat = 600 kcal

Useful Definitions

- *10% rule:* The amount of energy at each trophic level decreases as it moves through an ecosystem. As little as 10% of the energy at any trophic level is transferred to the next level; the rest is lost largely through metabolic processes (such as cellular respiration) as heat.
- *Calorie (kcal):* A unit of measurement of energy in food.

Calculations

Calculate the total area of the map in square miles.

Calculate the usable area of the map where deer could live (nonurban, open land).

Calculate the amount of grass available in the usable area.

Calculate the number of deer that can live on the amount of grass available.
(*Hint:* This can be a decimal.)

Calculate the number of mountain lions that can survive on the number of deer.
(*Hint:* This can be a decimal.)

Construct one or more graphs of your data comparing the amount of grass and the number of deer
and mountain lions that your area can support.

Do your results support the 10% rule? (Show your work!)

Some Useful Ideas From My Teacher

You can keep track of useful ideas from your teacher in the space below.

What We Figured Out

Now that you have completed this task, take a few minutes to fill out the Summary Table below with the other students in your group. This table will help you keep track of what you figured out during the task. You will then have an opportunity to share your ideas as we fill out our class Summary Table.

Summary Table

What we learned	How we know

How it helps us explain the phenomenon

Midunit Model Revision for Unit 2

Purpose

Now that your class is roughly halfway through the unit, it is important for students to go back to the initial models constructed on the first day and revise them based on what they have learned so far. Students may choose between two different strategies for model revisions. After negotiating what should be added, revised, or removed from their models, the groups may do either of the following:

- Redraw their models.
- Use sticky notes to keep track of their revisions for the final model revision near the end of the unit.

They should make the choice between these options based on the amount of time available for the model revision process. It is important, however, that students make decisions about revisions based on what fits with the evidence from the tasks or is consistent with the scientific ideas at play. We have included a model revision PowerPoint template for your use, which can be downloaded from the book's Extras page at *www.nsta.org/mbi-biology*, and Figure 2.9 contains resources to help with modeling.

Figure 2.9. Resources to Help With Modeling

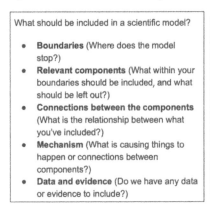
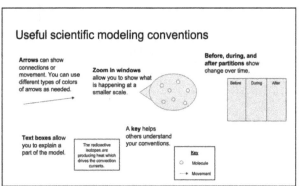

Although it does take time, we recommend a share-out session after the model revision. This allows for new ideas to be shared across the whole class, as well as discussion about what should and should not be included in the models. Have the members of each group stand up and describe how and why they changed their model. If time is limited, you could ask each group to focus on just one change. Provide time for questions from the class, and be sure to ask clarifying questions while comparing and contrasting the ideas presented across the groups.

Teacher Notes

Task 13. Trophic Cascades

Purpose

In the previous task, students learned about how energy flows through an ecosystem in the food web. In this task, students continue to examine interactions in food webs. However, here they focus on how one organism can directly or indirectly affect other organisms in an ecosystem, causing changes in their abundance through trophic cascades. We have adapted the "Exploring Trophic Cascades" activity from HHMI BioInteractive for this task. In the next task, students will extend this knowledge to examine how the living and nonliving parts of the ecosystem interact to cycle materials in the environment.

Important Life Science Content

Trophic cascades were described by Robert Paine (1980), who studied the effects of removing sea stars from the rocky shore in the Pacific Northwest. His observations led to the understanding that changes in abundance of one species can have far-reaching effects on many species in an ecosystem. This occurs because of the "cascades" of interactions. In the kelp forest ecosystem, for example, the removal of sea otters due to hunting had significant impacts on the entire ecosystem. The sea otters are predators of sea urchins. When they were removed from the ecosystem, the sea urchin population exploded. This is an example of a direct negative effect. The increased sea urchin population led to an increase in the amount of grazing on the kelp, another negative direct effect. As a result of these interactions, the sea otters' removal had an indirect positive effect on the kelp population. The removal of the sea otters further affected other species as well. The glaucous-winged gulls' primary food source switched from fish to invertebrates, and bald eagles became more heavily dependent on birds as a food source. The decline of sea otters in the kelp forest set off a trophic cascade, as their removal affected the abundance of organisms and feeding interactions throughout the ecosystem.

Scientific Ideas That Are Important to Think About During This Task

- Changes in populations of organisms within a food web can have important downstream impacts on other organisms in the ecosystem.
- Trophic cascades are powerful interactions that strongly regulate biodiversity and ecosystem function.

Timeline

Approximately one class period.

Materials and Preparation

The items needed for this investigation are listed in Table 2.9 (p. 206).

Table 2.9. Required Materials for Task 13

Item	Quantity
Computer or tablet with internet connection	1 per group
Student Handout	1 per student

Safety Precautions

This task does not require any specific safety precautions.

Procedure

This lesson plan is only a suggestion. It is included here to illustrate how you can facilitate student thinking during this task. We encourage you to modify this lesson plan by asking different questions, using different examples, and providing different scaffolds as appropriate to better meet the needs of students in your class.

Putting Ideas on the Table (50 minutes)

1. Before students begin the task, use the introductory slides and video on the website below to explore the idea of trophic cascades and provide directions for the task.

2. Working in their groups, students should go to the HHMI BioInteractive activity "Exploring Trophic Cascades" at *https://media.hhmi.org/biointeractive/click/trophiccascades*.

3. Groups should have discussions as they answer questions in the three parts of the website, keeping the idea of trophic cascades in mind throughout.

4. Conduct a whole-class discussion about trophic cascades as shown through the examples on the website.

Adding Information to the Summary Table (10 minutes)

1. Give students 5 minutes to decide what to add to the Summary Table at the end of the Student Handout.

2. Have one student from each group share what the group figured out, how they know (their evidence for what they figured out), and how this information will help them explain the anchoring phenomenon.

3. Once each group has shared, ask the entire class to decide what should be added to each column of the class Summary Table for Task 13. Help students reach consensus about what to add to the Summary Table. Only add an idea to the Summary Table if everyone in the class agrees with that idea.

Back Pocket Questions

As students work in groups, it is important to engage with each group to help press and extend students' thinking around the ideas at play in this task. Following are some example questions you might ask:

1. Helping students get started: What would you expect to see if there is a positive effect? If there is a negative effect?

2. Pressing further: What kinds of patterns are you seeing across the case studies?

3. Following up: What makes you think that? Can you say more?

Filling Out the Summary Table

Table 2.10 includes examples of the responses students may come up with when they fill out the Summary Table. This is provided here only as a behind-the-scenes roadmap and is not meant to be shared with students.

Table 2.10. Example Summary Table for Task 13

What we learned	How it helps us explain the phenomenon
Trophic cascades result from removing or adding an organism to an ecosystem, leading to positive or negative direct or indirect effects that have far-reaching consequences in an ecosystem.	Whale hunting indirectly affects the number of fish in the ocean. The presence of fewer whales leads to a decrease in fecal plumes, decreasing the phytoplankton, decreasing krill and zooplankton, decreasing fish.

Hints for Implementing This Task

- Draw a graph with a positive slope and one with a negative slope on the board as a reference for students when they are trying to determine whether the interactions in the case studies are positive or negative.

- The idea of positive and negative effects can be confusing for students because the words *positive* and *negative* do not necessarily correlate with an increase or decrease in population size. Therefore, it is important to establish the meaning of these terms before students begin this task. Positive effects occur when the change in one population leads to a similar change in another population. Examples of positive effects: a decrease in seabirds corresponds to a decrease in soil nutrients in the arctic tundra; an increase in phytoplankton in a lake corresponds to an increase in carbon influx. Negative effects, on the other hand, lead to changes in the opposite direction, such that an increase in one organism leads to a decrease in another. Examples of negative effects: an increase in the minnow population leads to a decrease in the phytoplankton population in a lake; an increase in the sea urchin population causes a decrease in the kelp population.

Possible Extensions

- The PBS Nature episode "The Serengeti Rules," available at *www.pbs.org/video/the-serengeti-rules-41dfru*, describes the trophic cascade in the African savanna and is a good supplemental resource for discussing this topic.
- Students can research other examples of trophic cascades to further investigate how one organism can affect many others in an ecosystem.

Reference

Paine, R. T. 1980. Food webs: linkage, interaction strength and community infrastructure. *Journal of Animal Ecology* 49 (3): 667–685.

Student Handout

TASK 13. TROPHIC CASCADES

Introduction

Trophic cascades reveal how one organism can directly or indirectly affect other organisms in an ecosystem, causing changes in their abundance. These effects can be far-reaching. The effects of a cascade can be direct or indirect, as well as positive or negative. *Direct effects* have immediate relationships to one another. *Indirect effects* are the subsequent events in a chain reaction. *Positive effects* occur when an increase in one factor leads to an increase in another factor or when a decrease in one factor leads to a decrease in another factor. *Negative effects*, on the other hand, result in the opposite occurring: an increase in one factor leads to a decrease in another factor, or a decrease in one factor leads to an increase in another factor. Trophic cascades are vitally important for understanding ecosystems, and scientists have documented them in ecosystems around the world.

Initial Ideas

Before beginning this task, take a few minutes to think about trophic cascades. How do trophic cascades affect organisms in an ecosystem?

How do you think changes in whale populations could cause a trophic cascade in the marine ecosystem?

Your Task

Your task is to continue to focus on interactions in food webs, specifically on how removal of essential organisms in a web can cause a trophic cascade. To accomplish this task, visit *https://media.hhmi.org/biointeractive/click/trophiccascades* and click on Introduction. Click through the introductory slides and watch the video to start learning more about the trophic cascades involved with the sea otters and kelp forests in the Pacific Ocean. After completing the section below, click on the Kelp Example and then the Case Studies tabs and answer the questions in the appropriate sections below. In the next task, you will investigate how whaling causes a trophic cascade that extends from the marine ecosystem into the atmosphere.

Introduction

Watch the video. Record what you notice and wonder about *how otters affect their ecosystem.*

What do you **notice**?	What do you **wonder**?

Kelp Example

1. What does a solid line represent?

2. What does the minus sign mean?

3. What does the dashed arrow represent?

4. Explain the interactions among the sea otters, sea urchins, and kelp in the kelp forest.

5. Examine the graphs for the gulls. Describe what happened to their diet when otters disappeared.

6. Examine the graphs for the bald eagle. Describe what happened to their diet when otters disappeared.

7. Explain WHY a change in the sea otter population in the kelp forest ecosystem is an example of a trophic cascade (*similar to a chain reaction*).

Case Studies

Click on each of the case studies listed below and complete the interactives to answer the following questions.

Arctic Tundra

8. Seabirds have a _____ (positive/negative) effect on soil nutrients. Explain your answer below.

9. Soil nutrients have a _____ (positive/negative) effect on the grasses. Explain your answer below.

10. The introduction of foxes had a _____ (positive/negative) and a/an _____ (direct/indirect) effect on the grasses in the Aleutian Archipelago. Explain your answer below.

Midwestern Lake

11. Bass have a _____ (positive/negative) effect on the minnows. Explain your answer below.

12. Minnows have a _____ (positive/negative) effect on zooplankton (microscopic animals). Explain your answer below.

13. Phytoplankton have a _____ (positive/negative) effect on carbon influx. Explain your answer below.

14. Explain how predatory bass affected the amount of carbon dioxide absorbed by the lake.

Some Useful Ideas From My Teacher

You can keep track of useful ideas from your teacher in the space below.

What We Figured Out

Now that you have completed this task, take a few minutes to fill out the Summary Table below with the other students in your group. This table will help you keep track of what you figured out during the task. You will then have an opportunity to share your ideas as we fill out our class Summary Table.

Summary Table

What we learned	How we know

How it helps us explain the phenomenon

Teacher Notes

Task 14. Carbon Cycle Game

Purpose

In the previous task, students examined trophic cascades in different ecosystems. This gave them the opportunity to see how one organism can affect many others. In this task, students extend their knowledge of trophic cascades to the carbon cycle. They figure out how an increase in algae in the ocean may, in turn, affect the amount of carbon dioxide (CO_2) in the atmosphere. Students "become" a carbon atom and simulate the path they may take throughout a variety of carbon reservoirs. They then create a diagram that shows how carbon moves throughout the environment. Students apply their knowledge of the carbon cycle to climate change and make connections to the driving question. In the next task, the students will conduct investigations to test their models.

Important Life Science Content

Matter cycles through inorganic and organic sources within and between ecosystems. Carbon, in particular, is "stored" in carbon reservoirs such as biomass, the atmosphere, or limestone. In a marine ecosystem, carbon is removed from the atmosphere by phytoplankton and other algae during the process of photosynthesis, which produces biological molecules necessary for growth. The carbon is transferred to different trophic levels as the phytoplankton are eaten. Most of the carbon is used to build the biomass of the organism that consumes it. The rest is released into the ocean as CO_2 through the process of cellular respiration. When organisms die, their remains can become part of the soil or sediment. This can lead to carbon sequestration.

Anthropogenic climate change is driven by fossil fuel combustion and the subsequent rise in atmospheric CO_2 concentrations. Carbon dioxide is a greenhouse gas that traps heat near Earth, leading to warmer global temperatures and climate change. The removal of whales from the oceans has affected the carbon cycle by removing nutrients used by phytoplankton. This has led to a decrease in phytoplankton populations and consequently a reduction in the amount of carbon dioxide removed from the atmosphere. This cascading effect has contributed to climate change.

Scientific Ideas That Are Important to Think About During This Task

- Carbon moves throughout the biosphere and geosphere through reservoirs. A number of different processes allow this movement.
- Humans have affected the amount of carbon in these reservoirs through our activities.

Timeline

Approximately two class periods.

Materials and Preparation

The items needed for this investigation are listed in Table 2.11. Place table signs at the appropriate table (or other location in the room). Box 2.1 (p. 217) describes the information that should be on the signs, which identify the carbon reservoir at each table.

Table 2.11. Required Materials for Task 14

Item	Quantity
Dice	2–3 per table
Table signs	1 per table
Computer or tablet with internet connection	1 per group
Student Handout	1 per student

Safety Precautions

This task does not require any specific safety precautions.

Procedure

This lesson plan is only a suggestion. It is included here to illustrate how you can facilitate student thinking during this task. We encourage you to modify this lesson plan by asking different questions, using different examples, and providing different scaffolds as appropriate to better meet the needs of students in your class.

Putting Ideas on the Table (110 minutes)

1. In Part 1, students play a game that is set up as a six-station activity across the classroom, with dice and signs at each table (see Box 2.1).

2. Introduce the activity using the PowerPoint template we have provided, which can be downloaded from the book's Extras page at *www.nsta.org/mbi-biology*.

3. Students should go through the stations in their groups, keeping track of their moves on the table.

4. Once groups have completed the required number of moves, students can return to their seats to discuss and answer the questions on the Student Handout.

5. In Part 2, students watch a video on climate change. This can be done as a whole class or within each group.

6. Students should complete the notice and wonder chart as a group.

7. Conduct a whole-class discussion on climate change and the carbon cycle game.

Adding Information to the Summary Table (10 minutes)

1. Give students 5 minutes to decide what to add to the Summary Table at the end of the Student Handout.

2. Have one student from each group share what the group figured out, how they know (their evidence for what they figured out), and how this information will help them explain the anchoring phenomenon.

3. Once each group has shared, ask the entire class to decide what should be added to each column of the class Summary Table for Task 14. Help students reach consensus about what to add to the Summary Table. Only add an idea to the Summary Table if everyone in the class agrees with that idea.

Box 2.1. Table Signs and Procedures for Each Station

Table 1: "CO_2 in the Atmosphere and Water"

- Roll 1 or 2: Plants and algae take up CO_2 during photosynthesis. Go to Table 6.
- Roll 3 or 4: CO_2 diffuses between air and water at the surface of lakes and oceans. Stay at Table 1. Roll again.
- Roll 5 or 6: Corals and mollusks take carbon out of the water to build shells. Go to Table 3.

Table 2: "Carbon Stored in Fossil Fuels"

- Roll 1, 2, or 3: Combustion of fossil fuels. Go to Table 1.
- Roll 4, 5, or 6: Untapped oil reserve. Stay at Table 2. Roll again.

Table 3: "Carbon Stored in Rocks and Shells"

- Roll 1 or 2: Weathering and erosion. Go to Table 1.
- Roll 3 or 4: Rocks buried under layers of sediment. Stay at Table 3. Roll again.
- Roll 5 or 6: Buried rocks returned to magma. Volcanoes release CO_2 into the air. Go to Table 1.

Table 4: "Carbon in Soil"

- Roll 1 or 2: Decomposition and decay of dead organic matter by fungus and bacteria releases CO_2 into the air. Go to Table 1.
- Roll 3 or 4: Dead organic matter stays in the soil/sediment. Humus forms. Stay at Table 4. Roll again.
- Roll 5 or 6: Organic matter is buried for millions of years and forms fossil fuels. Go to Table 2.

Table 5: "Carbon Stored in Animals"

- Roll 1 or 2: Animals carry out respiration, which releases CO_2. Go to Table 1.
- Roll 3 or 4: Animals die and become part of the soil or sediment. Go to Table 4.
- Roll 5 or 6: Animals pass waste products, which returns the carbon to the soil or sediment. Go to Table 4.

Table 6: "Carbon Stored in Plants and Algae"

- Roll 1 or 2: Plants and algae are eaten by animals. Go to Table 5.
- Roll 3: Plants and algae carry out respiration, which releases CO_2. Go to Table 1.
- Roll 4: Forests are burned by people to clear land. Go to Table 1.
- Roll 5 or 6: Plants and algae die and become part of the soil or sediment. Go to Table 4.

Back Pocket Questions

As students work in groups, it is important to engage with each group to help press and extend students' thinking around the ideas at play in this task. Following are some example questions you might ask:

1. Helping students get started: As a carbon atom, where are you spending the most time?

2. Pressing further: How do you think this relates to a greater number of whales contributing to a decrease in carbon dioxide in the atmosphere?

3. Following up: What makes you think that? Can you say more?

Filling Out the Summary Table

Table 2.12 includes examples of the responses students may come up with when they fill out the Summary Table. This is provided here only as a behind-the-scenes roadmap and is not meant to be shared with students.

Table 2.12. Example Summary Table for Task 14

What we learned	How it helps us explain the phenomenon
Molecules of carbon cycle between the biotic and abiotic parts of the environment through carbon reservoirs via a number of different processes. Living things such as phytoplankton remove and store carbon from the atmosphere and thus have an impact on climate change.	Fewer whales mean less nitrogen available in the ocean, leading to less phytoplankton. Smaller populations of phytoplankton capture less CO_2 from the air.

Hints for Implementing This Task

- To make the task more gamelike, set up the table cards so that students have to flip them over to find out what happens after they roll.

- Put a sign at each table indicating the carbon reservoir that table represents. This will help students keep track of where the carbon is moving to on their handout.

- If a student gets stuck in a particular reservoir for more than two turns, tell them it's okay to roll until they get something else so that they can move to another table.

Possible Extensions

- Have students compare their carbon cycles so they can see that there is not just one direct path that carbon can take as it is cycled throughout the environment.

- Students can analyze data at *www.biointeractive.org/classroom-resources/trends-atmospheric-carbon-dioxide*, which shows how atmospheric CO_2 levels have changed over time as part of the discussion about climate change.

- Have students examine other nutrient cycles, such as that of phosphorus.

- Students can examine carbon sequestration in whale falls, another way that whales can affect the carbon cycle, by visiting the following websites:

    - *www.heirstoouroceans.com/blog-collection/2018/1/27wzgb1wsr1k2uwpcjvdohk83zydc7pr*

    - *www.npr.org/2019/09/13/760664122/what-happens-after-a-whale-dies*

    - *www.youtube.com/watch?time_continue=1&v=CZzQhiNQXxU&feature=emb_logo.*

Student Handout

TASK 14. CARBON CYCLE GAME

Introduction

Carbon moves between the living and nonliving parts of the environment through carbon reservoirs. The biosphere (living portion) includes all the things living on and around Earth. The geosphere (nonliving portion) consists of the atmosphere (air), hydrosphere (water), and lithosphere (rocks and soil).

Initial Ideas

Before beginning this task, take a few minutes to think about the carbon cycle. Is there a link between whale plumes and the movement of carbon throughout the environment? Explain your answer.

Your Task

Your task is to extend your knowledge of trophic cascades to the carbon cycle. To accomplish this task, you will examine how carbon moves between the living and nonliving parts of the environment through carbon reservoirs. You will then apply your knowledge of the carbon cycle to climate change and make connections to the driving question. In the next task, you will test your model of the phenomenon.

Part I

Choose a table at which to start. Record your starting location in the space provided above the chart on the following page. Roll the die and follow the directions on the corresponding card. *Before you leave each table,* be sure to record "What Happened to the Carbon?" in the chart. *Once you move to a new table,* record "Where's the Carbon Now?" in the chart. Travel around the tables for a total of 10 rolls.

Starting Location: Table _____

Roll #	What Happened to the carbon?	Go to table #	Where's the carbon now?
	Example: Animals carry out respiration, which releases CO_2.	1	In the air
1st roll			
2nd roll			
3rd roll			
4th roll			
5th roll			
6th roll			
7th roll			
8th roll			

Continued

Roll #	What Happened to the carbon?	Go to table #	Where's the carbon now?
9th roll			
10th roll			

Analysis

Where did the carbon most often go?

In the space below, create a model to show the path your carbon atom took. Record the six locations where carbon goes in separate boxes. Connect them with arrows. Write HOW the carbon moved from one box to another on the arrows. *Challenge:* Add a whale plume to your model and show how this affects the carbon cycle.

Part 2

As you watch the video at *https://ed.ted.com/lessons/climate-change-earth-s-giant-game-of-tetris-joss-fong*, record what you notice and wonder in the chart below. Keep in mind the following questions as you watch:

- How have humans altered the carbon cycle?
- What are some of the effects of climate change?
- How is the carbon cycle like a game of Tetris?

What do you **notice**?	What do you **wonder**?

Some Useful Ideas From My Teacher

You can keep track of useful ideas from your teacher in the space below.

What We Figured Out

Now that you have completed this task, take a few minutes to fill out the Summary Table below with the other students in your group. This table will help you keep track of what you figured out during the task. You will then have an opportunity to share your ideas as we fill out our class Summary Table.

Summary Table

What we learned	How we know

How it helps us explain the phenomenon

TASK 15. TESTING YOUR MODEL: *CHLORELLA* INVESTIGATION

Teacher Notes

Task 15. Testing Your Model: *Chlorella* Investigation

Purpose

In the previous task, students learned that there are several factors that limit phytoplankton growth and reproduction in the ocean. In this task, the students investigate one of these factors: nitrogen. Using Miracle-Gro fertilizer and *Chlorella* algae, students test their models to see what impact nitrogen availability has on phytoplankton (*Chlorella*) growth. The results of this experiment will give them evidence to strengthen their explanations.

Important Life Science Content

Nitrogen is important for phytoplankton because it is an essential component in the production of chlorophyll. Chlorophyll is the pigment responsible for absorbing energy from light in order for photosynthesis to occur. Therefore, nitrogen can be a limiting factor in phytoplankton growth and reproduction. Miracle-Gro is a fertilizer that contains nitrogen and can be used to model the effect of whale plumes as a fertilizer for phytoplankton in the ocean. Increasing the amount of nitrogen that is available to the *Chlorella* will result in increased growth and reproduction. This can be measured using a spectrophotometer, which determines the absorption of light by a sample of water containing phytoplankton. When there is more *Chlorella* present, more light will be absorbed by the sample. Students can use the evidence from this investigation to provide an evidenced-based explanation for their models.

Scientific Ideas That Are Important to Think About During This Task

- Specific factors that limit the growth of populations are referred to as limiting factors.
- Models are not only created and revised but also tested against the real world.

Timeline

Approximately four or five class periods.

Materials and Preparation

The items needed for this investigation are listed in Table 2.13 (p. 227). Two weeks before the investigation, set up a culture of *Chlorella*. This can be ordered from any biological supply company. Boost the growth of the *Chlorella* culture with Alga-Gro (this can be purchased from Carolina Biological Supply at *www.carolina.com/dehydrated-media-and-media-ingredients/alga-gro-fresh-water-medium-1-qt/153752.pr*).

Table 2.13. Required Materials for Task 15

Item	Quantity
Safety goggles, nonlatex apron, and vinyl or nitrile gloves	1 per student
Chlorella culture	1 per class
Alga-Gro	1 per class
Miracle-Gro	1 per class
Test tube	1 per student
Test tube rack	1 per group
100-ml graduated cylinder	1 per group
10-ml graduated cylinder	1 per group
Eyedropper	1 per group
Labeling tape	1 per group
Marking pen	1 per group
Spectrophotometer	1 per group
Cuvette	1 per group
Whiteboards (4 × 2 feet) or chart paper	1 per group
Student Handout	1 per student

Spectrophotometers should be set to warm up for 10 to 20 minutes before use. Preset the spectrophotometer wavelength to 470 nm and set it to measure absorbance. Have a cuvette filled with distilled water near the spectrophotometer to use as a blank. Instructions for using the spectrophotometer can be found at *www.wiredchemist.com/chemistry/instructional/laboratory-tutorials/ using-the-spectronic-20-spectrophotometer*.

Safety Precautions

- Wear sanitized indirectly vented chemical-splash goggles and chemical-resistant, nonlatex aprons and gloves during lab setup, hands-on activity, and cleanup.
- Handle all glassware with care. Make sure students know what to do in case of broken glassware.
- Use caution when working with electrical equipment. Keep away from water sources, which can cause shorts, fires, and shock hazards. Use only GFI-protected circuits.

TASK 15. TESTING YOUR MODEL: *CHLORELLA* INVESTIGATION

- Immediately wipe up any spilled water on the floor so it does not become a slip or fall hazard.
- Wash hands with soap and water when the activity is completed.

Procedure

This lesson plan is only a suggestion. It is included here to illustrate how you can facilitate student thinking during this task. We encourage you to modify this lesson plan by asking different questions, using different examples, and providing different scaffolds as appropriate to better meet the needs of students in your class.

Putting Ideas on the Table

1. At the beginning of class, ask students how they would test their conclusions that nitrogen affects phytoplankton growth. What can they use to represent whale fecal plumes? You can discuss how nitrogen helps build a chlorophyll molecule that algae will use to collect light energy to carry out photosynthesis. In this discussion, you should also explain how you will use a spectrophotometer to measure algae growth. The more algae, the more light will be absorbed by the spectrophotometer.

2. Students should then design their experiments. You may want to help guide their design, but make sure you do not tell them how to carry out the experiment or what amounts to use. This is something students should do with their group members. The designs will not be perfect. Ask them if they think that another group would be able to do the same experiment with their procedure. You can sign off on the experimental design as long as the experiments include appropriate safety actions to be taken based on potential hazards and resulting risks.

3. After students complete the experiment, have them design argumentation boards in groups. We use large whiteboards (4 × 2 feet) purchased and cut from shower board at a local hardware store. You can also use chart paper. It is helpful for students to be able to see all the presentations side by side to visually compare them. It is not suggested that students do a computer presentation for this reason.

4. You can have groups use a whole-class or round robin presentation. If your students are not yet proficient in small-group discussions, the whole-class model will be easier. Provide students with sentence starters (e.g., "I notice that in your model . . ." "Can you help explain why . . . ?") to help them formulate questions to ask the presenters.

5. Next, discuss with students the validity of their experiments. Prompt students to compare the designs of different groups and what they would do to improve their experiments.

6. Finally, have students write an investigation report, summarizing the experiment they carried out and their argument supporting their claim.

Adding Information to the Summary Table (10 minutes)

1. Give students 5 minutes to decide what to add to the Summary Table at the end of the Student Handout.

2. Have one student from each group share what the group figured out, how they know (their evidence for what they figured out), and how this information will help them explain the anchoring phenomenon.

3. Once each group has shared, ask the entire class to decide what should be added to each column of the class Summary Table for Task 15. Help students reach consensus about what to add to the Summary Table. Only add an idea to the Summary Table if everyone in the class agrees with that idea.

Back Pocket Questions

As students work in groups, it is important to engage with each group to help press and extend students' thinking around the ideas at play in this task. Following are some example questions you might ask:

1. Helping students get started: What limiting factors do you think are at play in your investigation? Why?

2. Pressing further: How do you think this relates to our phenomenon of whale plumes? What from your investigation is similar? What is different?

3. Following up: What makes you think that? Can you say more?

Filling Out the Summary Table

Table 2.14 includes examples of the responses students may come up with when they fill out the Summary Table. This is provided here only as a behind-the-scenes roadmap and is not meant to be shared with students.

Table 2.14. Example Summary Table for Task 15

What we learned	How it helps us explain the phenomenon
Increasing the amount of Miracle-Gro leads to an increase in phytoplankton growth and reproduction.	Since whale plumes contain similar ingredients to Miracle-Gro, we now have evidence (data) that shows that an increase in whale populations (and consequently plumes) will have a positive effect on phytoplankton growth in the ocean.

TASK 15. TESTING YOUR MODEL: *CHLORELLA* INVESTIGATION

Hints for Implementing This Task

- Allow students to test any of the limiting factors from the previous task when designing their experiments to test their models. The following materials can be used for the student investigation: fertilizers such as Miracle-Gro for nitrogen or Ironite for iron, baking soda (sodium bicarbonate) for carbon, and diatomaceous earth for silica. Light availability can be tested by blocking light using screens or paper to cover the test tube or container being used to grow the *Chlorella*.

- If spectrophotometers are not available, you can conduct the investigation by evaluating the color change of the water due to algae growth using colorimeters.

Possible Extensions

- Another way to conduct the lab is by constructing algae growth chambers using water bottles. Students can do a visual comparison to determine whether algae have grown. This lab can be accessed on the website of the journal *The Science Teacher* at *www.nsta.org/store/product_detail.aspx?id=10.2505/4/tst19_086_09_35*. (Access is free for NSTA members; there is a nominal fee for nonmembers.)

Student Handout

TASK 15. TESTING YOUR MODEL: *CHLORELLA* INVESTIGATION

Introduction

There are several factors that limit phytoplankton growth and reproduction in the ocean, one of which is nitrogen. Whale plumes contain nitrogen, a nutrient that helps phytoplankton grow.

Initial Ideas

Before beginning this task, take a few minutes to think about the effects of nitrogen on phytoplankton. How can you test the claim that nitrogen in whale plumes affects the growth of phytoplankton?

Your Task

Your task is to design and conduct investigations testing the effect of Miracle-Gro, a fertilizer containing nitrogen, on the growth of *Chlorella*, a type of freshwater algae that is similar to saltwater phytoplankton. *Chlorella* grows best using cool-white fluorescent lights placed 45 to 60 cm above the sample and a room temperature of 72°F (22°C). To gather quantitative data, you will be using a device called a spectrophotometer to measure light absorption, the amount of light absorbed by a sample at a specific wavelength. The spectrophotometer will be set to absorb blue wavelengths of light, the wavelength absorbed by the chlorophyll found in *Chlorella*. The higher the absorbance of the sample, the more *Chlorella* are in the container.

Available Materials

You and your group may use any of the following materials during this task:

- Indirectly vented chemical splash goggles, nonlatex apron, and vinyl or nitrile gloves (required)
- Test tubes
- Test tube rack
- Miracle-Gro
- 100-ml graduated cylinder

- 10-ml graduated cylinder
- Eyedropper
- *Chlorella* culture
- Cuvette
- Spectrophotometer
- Labeling tape
- Marking pen

TASK 15. TESTING YOUR MODEL: *CHLORELLA* INVESTIGATION

Safety Precautions

Follow all normal lab safety rules. In addition, be sure to take the following safety precautions:

- Wear sanitized indirectly vented chemical-splash goggles and chemical-resistant, nonlatex aprons and gloves during setup, hands-on activity, and cleanup.
- Handle all glassware with care. If glassware is broken, do not touch it. Inform your teacher.
- Use caution when working with electrical equipment. Keep away from water sources, which can cause shorts, fires, and shock hazards. Use only GFI-protected circuits.
- Immediately wipe up any spilled water on the floor so it does not become a slip or fall hazard.
- Wash hands with soap and water when the activity is completed.

Procedure

1. Using a 100-ml graduated cylinder, add 15 ml of Miracle-Gro solution to a test tube. Be sure to wash out your graduated cylinder each time you get a new solution of Miracle-Gro.
2. Gently mix the *Chlorella* solution and use an eyedropper to add 10 drops of *Chlorella* to the test tube. You can gently swirl or invert the tube to mix the contents.

Observations: Record what you notice about the contents of the test tube in the space below.

3. Add 8 ml of solution from your test tube containing *Chlorella* into a cuvette. This cuvette will be used to measure absorbance with a spectrophotometer.

4. Place the distilled water cuvette into the chamber and push the power button. The spectrophotometer will show "BL" on the screen, and it will begin to blink just before it reads out to "0." This is called blanking and should be done before you begin measuring the absorbance for each of your solutions.

5. Carefully remove the blank. Gently swirl or invert the tube with your sample to mix the contents. Insert your sample and read the absorbance. Record this in the space below:

Sample absorbance: _____ nm

6. Return the solution to the test tube when complete.

7. Complete an investigation proposal to show how you will test the question, Does fertilizer affect the growth of *Chlorella*? (See p. 67 in Unit 1 for the proposal form.) When your instructor has approved your design, you may carry out the experiment.

8. Once your group has finished collecting and analyzing data, prepare a whiteboard that you can use to share your initial argument.

9. Your group will then share your argument with others. The goal of the argumentation session is not to convince others that your argument is the best one; rather, the goal is to identify errors or instances of faulty reasoning in the arguments so these mistakes can be fixed. You will therefore also critique other students' arguments, evaluating the content of the claim, the quality of the evidence used to support the claim, and the strength of the justification of the evidence included in each argument.

10. Once you have completed your research, you need to prepare an investigation report consisting of three sections that provide answers to the following questions:

   • What question were you trying to answer and why?

   • What did you do during your investigation, and why did you conduct your investigation in this way?

   • What is your argument?

Some Useful Ideas From My Teacher

You can keep track of useful ideas from your teacher in the space below.

What We Figured Out

Now that you have completed this task, take a few minutes to fill out the Summary Table below with the other students in your group. This table will help you keep track of what you figured out during the task. You will then have an opportunity to share your ideas as we fill out our class Summary Table.

Summary Table

What we learned	How we know

How it helps us explain the phenomenon

Building Consensus Stage Summary

In the building consensus stage of an MBI unit (Figure 2.10), students finalize their group models, compare and contrast those models as a class as they work to reach a consensus, and co-construct a Gotta Have Checklist of the ideas and evidence that should be part of their final evidence-based explanations. Before beginning this stage, review the relevant sections in Chapters 1 and 2 for specifics. See Figures 2.3 and 2.4 (pp. 167–168) for examples of a Gotta Have Checklist and a final model for this unit. We have provided a PowerPoint template for this stage, which can be downloaded from the book's Extras page at *www.nsta.org/mbi-biology*.

Figure 2.10. Building Consensus Stage Summary

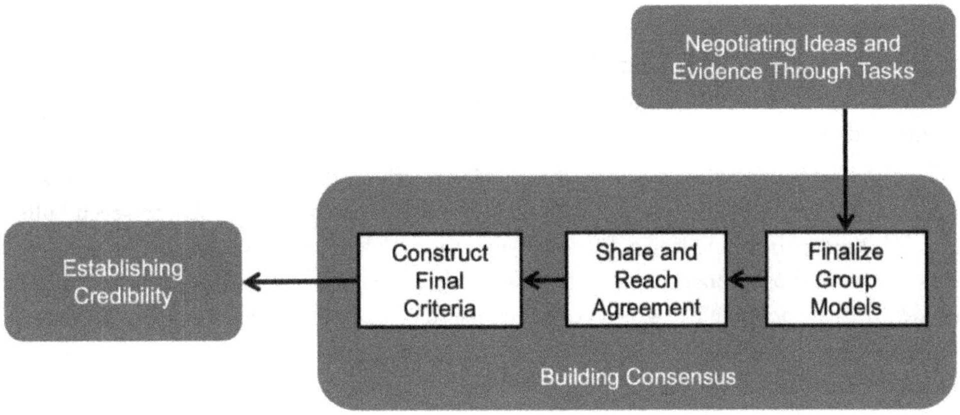

Finalize Group Models

Finalizing the models requires groups to review their previous models, decide what needs to be revised based on new ideas and understandings from the previous set of tasks, and redraw the models so they can be more easily shared and used to build consensus in a whole-class setting. This usually takes about 30 minutes. We often scaffold this process by asking students to review the completed Summary Table and talk about what should be added to, removed from, or changed in their previous models. They should think about what new ideas can help them move toward a truly causal model that includes not only *what* happened but also *why* it happened. This requires that the mechanism at play be visible and well explained. We push the groups to make sure that items in their models are labeled, that any unseen components are made visible, and that the important ideas that surfaced throughout the unit are explicitly used in the models.

Share and Reach Agreement

We think a share-out session is needed here so the groups can learn from each other and begin to build consensus across the models. There are a number of ways to do this, including gallery walks. As the goal is to build consensus as a whole class, we usually opt to facilitate a share-out session in which each group comes to the front of the class, displays its model, and talks it through. We prompt students to ask questions, and the class works hard to compare and contrast ideas across the models.

Construct a List of Final Criteria

The goal of our last public record, the Gotta Have Checklist, is to have the class negotiate about which of the main ideas and evidence should be part of a complete and scientifically defensible explanation of the phenomenon. Students will use this as a scaffold for writing their final evidence-based explanations in the last stage of the unit. There are a number of ways to facilitate this discussion and creation of the public record. We like to prompt groups to create a bulleted list of the three to five most important ideas that they think they need. We then ask for examples, press the whole class to make sure they understand how the idea fits in, and ask for consensus before writing it on the final checklist public record. This process can take about 15 to 20 minutes and most often goes quite smoothly by this point in the unit. However, this is also a time to make sure important ideas are included and to bring them up if necessary. While this is uncommon, we may also provide some just-in-time instruction to tie up any loose ends in students' understandings from the unit.

The checklist becomes less useful to students if everything that comes up in this discussion is automatically written on the board. Ideas can generally be combined into five to seven bulleted points. We then go back and lead a discussion about the evidence we have for each of the bulleted points and write those alongside. This step is crucial, as students will need to coordinate the science ideas with evidence in the written evidence-based explanation in the next stage. At times, we have been unhappy with our checklist for some reason; perhaps the writing is not as legible or the points are not as clearly articulated as we would like. In such cases, we created a clean version that evening and asked the students the next day to ensure that it still represented all the ideas from the original poster.

Establishing Credibility Stage Summary

In the establishing credibility stage (Figure 2.11), students construct arguments in the form of evidence-based explanations of the phenomenon, which allows them to argue their ideas in writing. This stage also includes peer review and revision to establish credibility for their ideas. Before beginning this stage, review the relevant sections in Chapters 1 and 2 for specifics. We have provided a PowerPoint template to assist with this stage, which can be downloaded from the book's Extras page at *www.nsta.org/mbi-biology*.

Figure 2.11. Establishing Credibility Stage Summary

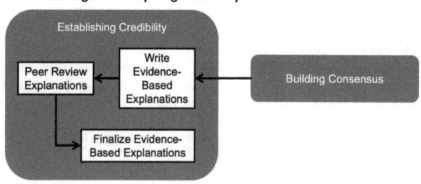

Write Evidence-Based Explanations

Up to this point, students have worked as part of a group to construct their evidence-based explanations of the phenomenon. Now, they work individually to write their final explanations. Although they are on their own, they have a number of important public records such as the Summary Table and Gotta Have Checklist to use as references. Beyond these scaffolds, you may choose to provide additional support for students who may struggle with writing. For instance, we often have the first sentence written out on small slips of paper that we can hand to students to help build momentum. Some teachers choose to have student groups work together to create a group outline of their writing before students work individually. Your school may have writing support structures in place to help students get started. One thing we have found is that teachers are most often pleasantly surprised by their students' individual evidence-based explanations.

Conduct Peer Review of Explanations

Once students have finished writing their individual evidence-based explanations, it is important that they receive peer feedback before turning in their final product. Have students use the MBI Explanation Peer-Review Guide in Appendix B to provide feedback to each other. Beyond helping students improve their written explanations, this step reinforces the idea that in science, credibility comes not from an external standard but through negotiation with our peers.

Finalize Evidence-Based Explanations

After students have been able to review their peer feedback, and perhaps meet with their peers to discuss the feedback, they should use the critique to finalize their evidence-based explanations.

Evaluation

Once students have handed in their written explanations, you can use the MBI Explanation Peer-Review Guide in Appendix B to evaluate them. Besides clarity of communication, the guide focuses on whether a student's final product explains the phenomenon, fits with the evidence gathered throughout the unit, and builds on the important science ideas at the center of the unit.

Unit 3. Heredity

Inheritance and Variation of Traits (LS3): Skin Cancer

Unit Summary

Cancer has affected nearly everyone, whether personally or through a family member or friend. In this unit, we use a specific case of skin cancer to explore the central dogma of biology: All cells contain genetic information in the form of DNA molecules that code for the formation of proteins, included in disciplinary core idea (DCI) Life Science 3 (LS3). Through the story of Freja Nicholson, students learn about skin cancer: the causes, who gets it, and how to recognize it. They investigate and compare normal and cancerous cells, examine the role of mutations in melanoma cases, and explore how mutations affect the functioning of proteins at the molecular level by learning about DNA structure and protein synthesis. Taken together, these powerful scientific ideas allow students to construct a scientific explanation of the case of Freja's skin cancer. Please note that no phenomenon can anchor all the ideas included in a single DCI. This unit does not fully cover ideas related to the variation of traits (LS3.B); these are included in Unit 4, on biological evolution, through studying the *Lampsilis* mussel.

Table 3.1 provides an outline for this unit, describing the purpose and the major tasks and products for each MBI stage, as well as an approximate timeline.

Table 3.1. MBI Outline and Timeline

MBI stage	Purpose	Major tasks and products	Timing
Eliciting ideas about the phenomenon	To introduce the phenomenon, eliciting students' initial ideas to explain the phenomenon, and begin constructing group models.	Initial ideas public record; Initial group models	1–2 days
Negotiating ideas and evidence through tasks	To provide opportunities for students to make sense of the phenomenon through purposeful tasks and discussions.	Individual task products; Revised group models	9 days
Building consensus	For groups to come to an agreement about the essential aspects of the explanation.	Final group models	1 day
Establishing credibility	For each individual student to write an evidence-based explanation.	Final evidence-based explanations	2 days
		Total	13–14 days

Anchoring Phenomenon

Freja Nicholson was a girl from England who enjoyed spending time at the beach and traveling with her family. Freja had a very light complexion and wore sunscreen when she traveled, but she often went without it when she was at home. She sometimes used tanning beds. At age 14, she had an atypical benign mole removed. Later, her doctors found a cancerous mole. The cancer eventually spread into her lymph nodes. Even though she had a number of lymph nodes removed, by the time she was 17, the cancer had spread to her breast, armpit, brain, and arm. Freja passed away at the age of 18. Her family has shared her story to help raise awareness of skin cancer in teenagers. Following are two articles that tell Freja's story:

- *www.today.com/health/mom-shares-story-daughter-s-skin-cancer-death-help-save-t100659*
- *www.mirror.co.uk/news/uk-news/girl-died-arms-kept-thinking-8282928*

Driving Question

What caused Freja's skin cancer?

Target Explanation

Following is an example evidence-based explanation that could be expected at the end of the unit. We consider this an exemplar final explanation at this grade level. This is included here to help support you, the teacher, in responsively supporting students in negotiating similar explanations by the end of the unit. It is not something to be shared with students but is only provided as a behind-the-scenes roadmap for you to consider before and throughout the unit to guide your instruction.

Freja Nicholson, an 18-year-old from England, died from melanoma, a form of skin cancer. Skin cancer is influenced by a variety of factors: pigmentation of the skin, a weakened immune system, exposure to the Sun, family history, and where you live (distance from the equator and elevation). Freja spent a good deal of time during her youth outdoors, and although she used sunscreen when she went on vacation, she did not often use it when at home in England. This left her skin exposed to damaging UV radiation from the Sun. In addition, Freja, who had a light skin complexion, used tanning beds, increasing her exposure to UV radiation. In the What Is Melanoma? task, we learned that repeated exposure to UV radiation from the Sun can lead to damage in the DNA of skin cells. Therefore, it is likely that Freja's sun exposure during her childhood is what led to the development of melanoma. In the Freja's Factors task, we learned that one of the first signs of melanoma is a change in existing moles or new pigmentation or growths on the skin. Freja's mother had noticed changes to a mole on her back, which was removed by doctors and found to be benign. Her doctor said that it was likely that a different mole on her body became cancerous without her knowledge.

Mutations affecting the cell cycle cause the most detrimental effects to the skin. Throughout the cell cycle, there are checkpoints at which the cell receives signals to continue to divide, to rest, or to stop and repair. In the Control of the Cell Cycle Online Simulation task, we learned that if the proteins regulating the cell cycle are altered, the cells will divide uncontrollably because they are not getting the signal to stop dividing. Most of the time, cells are able to repair the mutations, but if the mutation is not repaired and is in a location that affects cell growth and division, uncontrolled cell growth may occur and produce cancerous tumors.

In the DNA Structure Pop Bead Model task, we discovered that DNA is built from four different nucleotides: thymine, adenine, guanine, and cytosine. The sequence of nucleotides in a gene provides a code for a protein with a unique structure and function. In the What Are Proteins? task, we learned that the proteins that provide signals that control the cell cycle are created based on the specific genetic code in a person's DNA. If a piece of the original DNA sequence is altered in any way, the resulting protein may no longer function to provide the proper signal in the cell cycle. If tumor suppressor (inhibitor) proteins, such as p53, are not present or are not functional, the cell will continue to grow and divide. Additionally, if oncogenes are altered, they could give the cell the incorrect signal to continue to grow and divide, even though it is not ready. The mutations in Freja's cells were most likely caused by exposure to UV radiation that altered the DNA code in these genes, causing them to produce nonfunctional inhibitor or START proteins, which led to uncontrolled cell growth and division.

In order to function, cells need to turn instructions encoded in their DNA into proteins. In the Protein Synthesis Simulation task, we learned that for a cell to build the protein, the DNA is first transcribed into messenger RNA (mRNA). The mRNA delivers the transcribed code to the ribosomes, where it is translated into a chain of amino acids that have been delivered by the transfer RNA (tRNA) molecules. These amino acids are released into the cell, where they fold into a unique shape that will determine their function. In the Mistakes Happen task, we discovered that if a piece of DNA is mutated, it will not provide the same information to create the same protein. The resulting shape of the protein will be altered, possibly making the protein nonfunctional. Since every protein serves a specific function within cells and our bodies, minor changes in DNA can have large-scale effects, as altered proteins often serve a detrimental purpose. Since UV radiation damaged the DNA in Freja's skin cell, that cell was unable to produce functional tumor suppressor and proto-oncogene proteins involved in the cell cycle. This led to loss of control of the cell cycle and excessive cell division, resulting ultimately in the tumors that caused her death.

Eliciting Ideas About the Phenomenon

The eliciting ideas about the phenomenon stage of MBI is about introducing the phenomenon and driving question, eliciting students' initial ideas about what might explain the phenomenon and answer the driving question, and constructing their initial models in small groups. We have provided a PowerPoint for this purpose that will lead you step-by-step through the process, which can be downloaded from the book's Extras page at *www.nsta.org/mbi-biology*. There are two products of this stage: an Initial Hypotheses List and student groups' initial models. Examples of each are shown in Figures 3.1 and 3.2 (p. 242).

Figure 3.1. Example Initial Hypotheses List

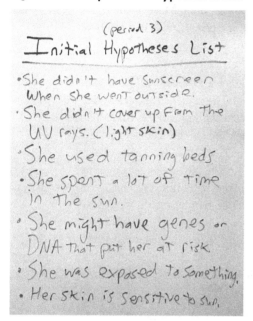

This stage of MBI is critically important, as it orients students to the MBI unit anchoring phenomenon using the driving question in a way that makes it compelling enough to spend several weeks across a unit working on explaining an event. Put more succinctly, this stage introduces the problem space in which the students will be working. Introducing the phenomenon should be done in a way that supports students in drawing on what they have learned previously in and outside of school as they begin to think about how to explain the driving question. Your priority should be to elicit students' ideas about the phenomenon, since it is important for them to continually think about the phenomenon and refine their everyday ways of thinking. Since eliciting students' ideas is a top priority, it's important to create a learning environment where students feel comfortable and are invited to offer ideas. This means that you need to think carefully about how to get students to float (or put on the table for consideration) as many ideas as possible. At this stage, there are no right or wrong ideas. Everything (within reason!) is on the table.

Figure 3.2. Example Initial Model

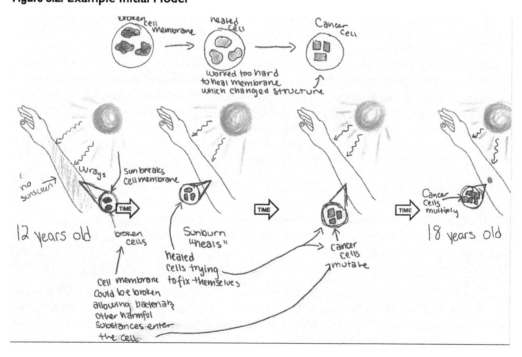

Negotiating Ideas and Evidence Through Tasks

The negotiating ideas and evidence through tasks stage of MBI makes up the majority of the unit. Each task is designed to introduce a key scientific idea or to reinforce an idea already raised by students in the class. Halfway through the unit, we recommend having groups revise their models and share out those revisions. PowerPoint templates are provided for the tasks as well as for the model revision; these can be downloaded from the book's Extras page at *www.nsta.org/mbi-biology*. Each task begins by introducing ideas for students to reason with while working on the task. During the task, students investigate or test the concepts using data. We recommend using the Back Pocket Questions provided in the Teacher Notes to press students' thinking while they are engaged in the task. Taken together, the tasks scaffold students' thinking as they co-construct a scientific explanation of the phenomenon. The general outline of this stage is as follows:

Task 16. What Is Melanoma?

Task 17. Freja's Factors

Task 18. Control of the Cell Cycle Online Simulation

Midunit Model Revision for Unit 3

Task 19. DNA Structure Pop Bead Model

Task 20. What Are Proteins?

Task 21. Protein Synthesis Simulation

Task 22. Mistakes Happen

A Summary Table is used after each task to scaffold students' negotiation of ideas and how those ideas help explain the phenomenon. An example is provided in Table 3.2 (p. 244). Like the target explanation, this is included here to serve as a behind-the-scenes roadmap to help support you, the teacher, in responsively supporting students in negotiating similar responses by the end of the unit. It is not something to be shared with students but is provided for you to consider before each task to guide your instruction.

Table 3.2. Example Summary Table for Unit 3

Task	What we learned from this task	How it helps us explain the anchoring phenomenon
16. What Is Melanoma?	Skin cancer is a disease resulting from an uncontrolled growth originating in the epidermis of skin cells. Breaks (mutations) in the DNA cause uncontrolled cell growth. Skin cancer may be detected by looking for irregularity in moles on the skin.	Freja was overexposed to UV light, had a light complexion, and developed atypical moles, all of which increased her chances of acquiring melanoma.
17. Freja's Factors	There are several risk factors that contribute to melanoma. These include increased age, use of tanning beds, geographic location (Europe, Australia, North America), light skin color, and lack of sunscreen use.	Freja had several risk factors that likely contributed to her melanoma. She had light skin and lived in England. She had used tanning beds and often did not use sunscreen while at home. Her age was not a risk factor that contributed to her melanoma.
18. Control of the Cell Cycle Online Simulation	Normal cells go through several checkpoints in the cell cycle that regulate the timing of cell division. Cancerous cells are not regulated by the checkpoints and continuously move through the cell cycle without stopping. This leads to the overproduction of cells that eventually form a tumor.	Cancerous cells are not regulated by the cell cycle checkpoints and divide continuously. Freja's cancerous cells were dividing out of control and formed tumors.
19. DNA Structure Pop Bead Model	DNA is a molecule consisting of two chains. Attached to the backbone of each chain are nitrogen bases (A, T, C, G). Each nitrogen base is bonded together with a complementary base (A-T and C-G). The structure of DNA allows it to be copied as the complementary bases pair up.	This helps us understand the directions in the DNA of cells and what the mutations might affect. The genes involved in Freja's cancer were coded by the DNA nucleotide sequences.
20. What Are Proteins?	The sequence of nucleotides determines the sequence of amino acids in a protein. Each protein serves a specific function within an organism.	The proteins in Freja's skin cells were made using the directions in her genes. Specific proteins, such as p53, may be important in producing cells that divide properly.
21. Protein Synthesis Simulation	DNA is transcribed into mRNA, which is translated into amino acids to create different proteins. The sequence of DNA nucleotides on a gene affects the resulting protein.	The proteins in Freja's skin cells were made using the particular nucleotide sequence of her genes. If the DNA had been altered by UV radiation, different proteins would have resulted.

Continued

Table 3.2. Example Summary Table for Unit 3 (continued)

Task	What we learned from this task	How it helps us explain the anchoring phenomenon
22. Mistakes Happen	Mutations are caused by a change in the sequence of DNA nucleotides that affects the mRNA nucleotides produced and the amino acids created. The sequence of amino acids affects the shape of the resulting protein.	The mutations that occur in areas responsible for normal functioning of the cell cycle lead to a change in the sequence of the tumor suppressor gene or oncogene, causing mRNA, amino acid sequences, and finally the shapes of proteins to be altered. The proteins can no longer do the same job, resulting in uncontrolled cell division.

You can think of your role here as that of helping introduce ideas to students to support them in learning about and beginning to engage with the ideas through carefully designed sensemaking tasks. If important ideas are not introduced by the students, you can put them on the table or introduce them through the use of just-in-time direct instruction, short videos, or readings, among other strategies. Since the objective is to have students pick up and try out these ideas in carefully crafted tasks as they develop explanations, introducing your students to an idea early in the task is *not* giving them answers that will be confirmed in the task. Instead, you can think of this as introducing students to tools (i.e., disciplinary core ideas) that will be useful in a task in ways that will help them make sense of data, a simulation, an activity, or something in the world. In the end, it is important that you think of introducing new ideas in the early parts of this MBI stage as an opportunity for students to engage in sensemaking with new ideas to use, while you concurrently recognize the potential explanatory power of the new ideas. At the end of each task, the completion of a row of the Summary Table is an opportunity for students to think about what they have learned in the task and, together with peers, reason about how the newly introduced ideas can be applied to support their objective of explaining the unit anchoring phenomenon.

Building Consensus

In the building consensus stage of an MBI unit, the whole class works to build consensus about the explanation of the phenomenon by finalizing the groups' models, comparing and contrasting those models as a whole class, and constructing a consensus checklist of the ideas and evidence (called the Gotta Have Checklist) that should be a part of students' final evidence-based explanations that make up the summative assessment of the unit. We have provided a PowerPoint as a guide through this stage of the unit, which can be downloaded from the book's Extras page at *www.nsta.org/mbi-biology*. Examples of the whole-class Gotta Have Checklist and a group's final model appear in Figures 3.3 and 3.4 (p. 246).

Figure 3.3. Example Gotta Have Checklist

IB ONLY SKIN DEEP? GOTTA HAVE ✓LIST

-KATIE LIKELY INHERITED A MUTATED
ALLELE FROM 1 OF HER PARENTS ON THE
GENE CODING FOR TUMOR SUPPRESSOR PROTEINS
(p53, ATM, BRCA1) WHICH NORMALLY STOP CELL
@ CHECKPOINTS
-DUE TO LOW MELANIN CONTENT IN HER SKIN CELLS,
KATIE HAD LESS PROTECTION OF HER DNA FROM UU
RAY
-UU RADIATION DAMAGED THE PART OF HER (CONTINUED TO DAMAGE)
DNA CODING FOR CHECKPOINT PROTEINS → THUS
MUTATING THE OTHER TUMOR SUPPRESSING ALLELE
-THE MUTATION LED TO SYNTHESIS OF NONFUNCTIONAL
p53, ATM, BRCA1 → CELLS REPLICATE MUTATED
DNA + KEEP REPLICATING WHEN THEY SHOULD STOP

In Chapter 2, we described the scientific practice of modeling as a knowledge-building practice and models as the products of modeling. Students begin this stage by finalizing their models in groups of three or four, which they should think of as a continuation of the sensemaking that started in the groups' initial modeling experiences in the early days of the MBI unit and continued when they revisited these models midway through the unit. As with these other experiences, you can expect the practice of modeling to support groups in refining their explanations of the unit anchoring phenomenon. Here, negotiation and argumentation lead to refined ways of thinking about the phenomenon that may not have occurred had students not engaged in modeling this final time. The groups'

Figure 3.4. Example Final Model

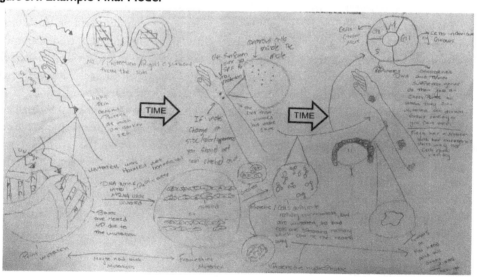

models will offer insight into your students' thinking that otherwise may not be accessible to you. At the same time, models as products of modeling also serve as artifacts that students can compare across groups in ways that are supported by your selection and sequencing of groups' sharing in whole-class sensemaking, so that differences can be foregrounded, negotiated, and resolved through argumentation, among other science practices.

Finally, the construction of final criteria or whole-class mapping of ideas included in the Gotta Have Checklist to the evidence collected from the various tasks completed in the second stage of MBI is a scaffold that will be important for individual students in writing evidence-based explanations. Importantly, the teacher-facilitated, co-constructed Gotta Have Checklist outlines what the class agrees should be included in students' explanations and is mapped to evidence identified across the unit tasks. This step is the last stage of building consensus. This is especially true because final group models will have been negotiated in both small and whole groups as a basis for the identification of the criteria that are needed for writing an evidence-based explanation.

Establishing Credibility

In MBI, students must argue for their ideas in writing. In the establishing credibility stage, they do this through written evidence-based explanations, peer review, and revision. We have provided a PowerPoint template as a guide through this stage of the unit, which can be downloaded from the book's Extras page at *www.nsta.org/mbi-biology*. Writing scaffolds such as sentence stems may be necessary to jump-start some students' writing. Once the written explanations are complete, we recommend using the MBI Explanation Peer-Review Guide found in Appendix B. The target explanation on pages 240–241 is an example of a written explanation.

Writing evidence-based explanations is a sensemaking experience that students engage in early in this stage of MBI. This is another knowledge-building opportunity that, like engaging in modeling earlier in the unit, is a practice that leads to a product. And like models, which served as products that could provide you with insight into the way groups were thinking about explaining the anchoring phenomenon, the individually written evidence-based explanations afford you with an opportunity to assess individual students in terms of where they are in their attempts to explain the anchoring phenomenon. As students engage with others in this stage, this is an additional space for argumentation and negotiation as they recognize possible differences and details they may not have considered or may not yet agree with.

This practice of peer review needs to be supported, so we have provided the MBI Explanation Peer-Review Guide in Appendix B, which students can use to provide feedback to each other. The peer-review guide contains questions designed to focus student attention on the most important aspects of explanations and to encourage students not only to discuss what they figured out and how they know what they know but also to best communicate what they have learned to others in a way that is clear, complete, and persuasive. During the peer review, you should consider and make explicit how explanations, like models, are refined over time through negotiation within a community. The MBI unit culminates at the end of this stage as students use the feedback they have received to make final revisions that represent how they are thinking about the unit anchoring phenomenon. The final evidence-based explanations serve as one measure of what students have learned about engaging in science practices to use disciplinary core ideas and crosscutting concepts to explain the phenomenon.

Hints for Implementing the Unit

- Students have different experiences with cancer. Some students may be sensitive to this issue, so it is important to begin the unit with a discussion about how to be respectful toward one another.
- We recommend providing opportunities to connect students' personal experiences and community resources with the phenomenon and topic of this unit. Even though students may

not have had any personal experiences with the phenomenon, we suggest holding discussions throughout the unit to elicit their personal connections. Think about possible connections in your community that you can bring to the conversation as well. These kinds of connections will make the unit personally meaningful to your students, increasing their motivation to engage in the tasks. In our experience, much of this happens during the eliciting ideas about the phenomenon stage of the unit. However, you should prompt students to connect to the topic during all phases of the unit.

- There may be times during the unit when students make connections to injustices or disparities in their communities or raise emotional responses to the topics. We suggest preparing to address young people's questions and desire for activism so that science practices are not portrayed as disconnected from the social and cultural contexts of students' real-world experiences.

Targeted NGSS Performance Expectations

- **HS-LS3-1.** Ask questions to clarify relationships about the role of DNA and chromosomes in coding the instructions for characteristic traits passed from parents to offspring.
- **HS-LS3-2.** Make and defend a claim based on evidence that inheritable genetic variations may result from: (1) new genetic combination through meiosis, (2) viable errors occurring during replication, and/or (3) mutations caused by environmental factors.
- **HS-LS3-3.** Apply concepts of statistics and probability to explain the variation and distribution of expressed traits in a population.

Eliciting Ideas About the Phenomenon Stage Summary

The first stage of MBI, eliciting ideas about the phenomenon (Figure 3.5), involves introducing the anchoring phenomenon and driving question, eliciting students' initial ideas and experiences that may help them develop initial explanations of the phenomenon, and developing initial models of the phenomenon based on those current ideas. Before beginning this stage, review the relevant sections in Chapters 1 and 2 for specifics. See Figure 3.1 on page 242 for an example of an Initial Hypotheses List for this unit. We have provided a PowerPoint template to assist with this stage, which can be downloaded from the book's Extras page at *www.nsta.org/mbi-biology*.

Figure 3.5. Eliciting Ideas About the Phenomenon Stage Summary

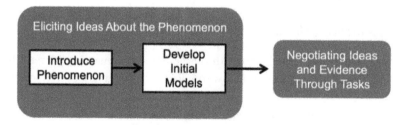

National Science Teaching Association

Introduce the Phenomenon

We begin this stage by introducing the phenomenon in an engaging way, such as with stories, videos, demonstrations, or even short activities. The goal is to provide just enough information for students to begin to reason about the phenomenon, without providing too much of the explanation. While we are introducing the phenomenon, we ask questions to keep students engaged and make sure they are paying attention to the important aspects that can begin to help them explain the phenomenon. The introduction ends with the driving question of the unit, which we have found helps focus their thinking on the development of a causal explanation for the phenomenon, or a "why" answer.

It is important to spend time eliciting not only students' scientific ideas about the phenomenon but also their personal connections to the phenomenon. What personal experiences do they have that help them connect to the phenomenon? While the phenomenon of the unit may not be directly connected to your community, does the community have similar phenomena or specific resources that you can bring into the discussion? The more connected students feel to the unit phenomenon, the more engaged and motivated they will be.

We then get students into their groups and facilitate the first discussion to try to answer the driving question with just the resources they brought with them—the ideas, experiences, and cultural resources they have gained both in and outside of the classroom. These ideas may be fully formed, partially correct, or fully incorrect in terms of our canonical knowledge of science. However, we make it clear that all ideas are considered equally valid at this point in time, as we realize these are the ideas that students put into play when they think about the phenomenon we have introduced.

Once student groups have discussed their ideas, we facilitate a class discussion to compare and contrast the ideas generated by each group. As ideas are presented, they are put on our first public record, which we call the Initial Hypotheses List. We use this list throughout the unit to keep track of the changes in students' thinking as they work toward a final evidence-based explanation for the anchoring phenomenon. We consider all ideas to be valid at this stage, before students use other resources to begin making sense of how or why something happens. The Initial Hypotheses List is also useful for the next task in this stage, initial model construction, especially since it offers students additional ideas beyond those they initially had either individually or in their small groups.

Develop Initial Models

If students are not experienced with modeling, it is worth providing a brief introduction and example. We have included an example in the PowerPoint for this stage, which can be downloaded from the book's Extras page at *www.nsta.org/mbi-biology*. Once the class is ready to begin modeling, we give each group a sheet of 11 × 17 inch paper and ask the groups to each make a model of their initial hypothesis. Sometimes they choose their own original hypotheses, and other times they are influenced by their peers' ideas and adopt one of them instead. As the groups work on constructing their models, you should walk around asking clarifying questions and pushing students to be as specific as possible. Once the models are ready, it is important to have students share ideas across them. There are a number of ways you might run these share-out sessions. We often collect and present the models on a document camera at the end of the first day. Groups can provide one- or two-sentence summaries of the initial hypotheses they have represented in their models. We point out interesting ideas and ways in which they have represented these ideas. For example, we may call attention to the fact that a group labeled the arrows, which made the model more understandable, and that another group used a zoom-in window to show what was happening at a different scale. At the end of the first day and this first stage, we have elicited ideas across the class, and the groups' initial hypotheses and models will act as a starting point for the rest of the unit.

Negotiating Ideas and Evidence Through Tasks Stage Summary

The next stage of MBI, negotiating ideas and evidence through tasks (Figure 3.6), takes up the majority of the unit. The tasks are designed to introduce or extend important science ideas that students need as they construct their evidence-based explanations of the phenomenon. Before beginning the tasks, we suggest you review the relevant sections in Chapters 1 and 2 for specifics. Each task consists of the following:

- A **Teacher Notes** section that provides an overview of the activity and science content important to the task. Here, we provide guidance on conducting the task with details on the procedure, a list of required materials and preparation, any necessary safety precautions, suggested Back Pocket Questions, and an example Summary Table entry. At the end of this section are further hints for implementing the task and possible extensions, if desired.

- A **Student Handout** to be given to groups as they engage in the task. The handout provides an introduction to the task and content, an initial ideas section to frame the learning before they begin, and detailed instructions and work space. It ends with a section called What We Figured Out, which is designed to scaffold student responses to include in the Summary Table as a whole class. We suggest having students fill out this section after completing the task and the post-task discussion.

- A **PowerPoint** Negotiating Ideas and Evidence Through Tasks template that you can adapt for each task, which can be downloaded from the book's Extras page at *www.nsta.org/mbi-biology*.

- A Midunit Model Revision reminder roughly halfway through the unit.

Figure 3.6. Negotiating Ideas and Evidence Through Tasks Stage Summary

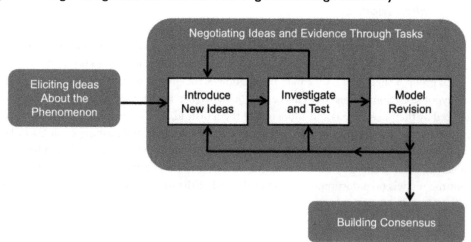

National Science Teaching Association

Teacher Notes

Task 16. What Is Melanoma?

Purpose

The purpose of this task is to have students become familiar with the different types of skin cancer, the causes of skin cancer, and ways to prevent skin cancer. In the introduction to the unit, students learned about Freja, a teenager from England who died from skin cancer. While it is likely that students know someone who has experienced cancer, they may not know details about what cancer is, how it develops, or who is at risk for developing it. Students also learn about how the cell cycle of cancerous cells is different from that of normal cells.

Important Life Science Content

The skin consists of three layers: the epidermis, dermis, and subcutaneous tissue. The outermost layer, the epidermis, contains melanin and keratin, which protect the deeper layers of skin. Exposure to ultraviolet (UV) radiation from the Sun causes damage to the DNA in skin cells. If the skin is subjected to repeated and prolonged exposure, the DNA will not be able to repair itself. The damaged DNA causes excessive cell division and will be replicated and passed on to daughter cells during mitosis. This leads to skin cancer. Benign and malignant tumors form as a result of excessive cell division. The difference between these two types of tumors is that benign tumors are typically not life threatening, do not spread to other parts of the body, and do not interfere with regular body functions, while malignant tumors may spread throughout the body, do interfere with regular body functions, and are life threatening. There are several risk factors that can make a person more susceptible to developing skin cancer: Being over the age of 40, having a light skin complexion, being exposed to high levels of UV radiation from the Sun, having a family history of skin cancer, and having actinic keratosis all increase the risk of developing skin cancer.

Melanoma is the most dangerous form of skin cancer, in part because it is fast-growing. It develops in the pigment-containing melanocytes within the epidermis of the skin. In addition to the risk factors mentioned previously, people who have birthmarks and freckles, have had three or more blistering sunburns, or have spent three or more years at an outdoor job are more susceptible to developing melanoma. Melanoma usually forms as a growth on the skin or a mole and can be identified by asymmetrical growth, irregular borders, multiple colors, or a large diameter (>6 cm).

People can reduce their risk of skin cancer by wearing protective items such as hats, sunglasses, and long-sleeved shirts and pants; applying sunscreen; or blocking the Sun with an umbrella. Finally, people should check their skin regularly and take note of any changes in moles.

Scientific Ideas That Are Important to Think About During This Task

- Cancer is simply uncontrolled cell growth. Skin cancer is the uncontrolled growth of melanocytes, cells in the epidermal layer of the skin.

Timeline

Approximately one class period.

Materials and Preparation

The items needed for this investigation are listed in Table 3.3. In preparation for this task, we recommend that you familiarize yourself with the websites students may use to answer the questions on the handout (several suggested websites are listed in the Procedure section).

Table 3.3. Required Materials for Task 16

Item	Quantity
Computer or tablet with internet connection	1 per group
Student Handout	1 per student

Safety Precautions

This task does not require any specific safety precautions.

Procedure

This lesson plan is only a suggestion. It is included here to illustrate how you can facilitate student thinking during this task. We encourage you to modify this lesson plan by asking different questions, using different examples, and providing different scaffolds as appropriate to better meet the needs of students in your class.

Putting Ideas on the Table (50 minutes)

1. Use the PowerPoint template we have provided, which can be downloaded from the book's Extras page at *www.nsta.org/mbi-biology*, to introduce skin cancer and give instructions for the task.

2. Ask students to begin working through the questions on the Student Handout. Students should work in groups on this task, discussing appropriate responses for the questions as they work together to find relevant information on skin cancer. While students are not provided a starting point for their investigation of skin cancer, the following websites would be appropriate if any groups are struggling to get started:

   - *www.cdc.gov/cancer/skin/basic_info/index.htm*
   - *www.mayoclinic.org/diseases-conditions/skin-cancer/symptoms-causes/syc-20377605*
   - *www.skincancer.org*

3. Rotate among the groups, asking appropriate Back Pocket Questions (see next page) to help students make sense of the content.

4. At the end of the task, conduct a whole-class discussion to aid student understanding of skin cancer.

Adding Information to the Summary Table (10 minutes)

1. Give students 5 minutes to decide what to add to the Summary Table at the end of the Student Handout.

2. Have one student from each group share what the group figured out, how they know (their evidence for what they figured out), and how this information will help them explain the anchoring phenomenon.

3. Once each group has shared, ask the entire class to decide what should be added to each column of the class Summary Table for Task 16. Help students reach consensus about what to add to the Summary Table. Only add an idea to the Summary Table if everyone in the class agrees with that idea.

Back Pocket Questions

As students work in groups, it is important to engage with each group to help press and extend students' thinking around the ideas at play in this task. Following are some example questions you might ask:

1. Helping students get started: What have you learned about skin cancer? Is this what you thought skin cancer was?

2. Pressing further: How does what you learned today relate to Freja's story? What parts are relevant?

3. Following up: What makes you think that? Can you say more?

Filling Out the Summary Table

Table 3.4 includes examples of the responses students may come up with when they fill out the Summary Table. This is provided here only as a behind-the-scenes roadmap and is not meant to be shared with students.

Table 3.4. Example Summary Table for Task 16

What we learned	How it helps us explain the phenomenon
Skin cancer is a disease resulting from an uncontrolled growth originating in the epidermis of skin cells. Breaks (mutations) in the DNA cause uncontrolled cell growth. Skin cancer may be detected by looking for irregularity in moles on the skin.	Freja was overexposed to UV light, had a light complexion, and developed atypical moles, all of which increased her chances of acquiring melanoma.

Hints for Implementing This Task

- Depending on the students' familiarity with searching for information on the internet, it may be necessary to provide a short introduction to finding expert health information (e.g., focusing on results from the CDC and other agencies).
- To reduce the time needed for this task, you can provide specific curated websites as needed.

Possible Extensions

- Students will have a variety of prior experiences with cancer, as well as differing levels of background knowledge about the basics of skin and skin color and exposure to UV radiation. The following extensions will help increase engagement and build background knowledge:
  - Have students listen to NPR's Code Switch podcast episode "Will Your Melanin Protect You From the Sun?" at *www.npr.org/sections/codeswitch/2018/07/05/559883985/will-your-melanin-protect-you-from-the-sun*.
  - Have students read the medical news article "Individuals of Color Have Poor Melanoma Survival" at *www.medpagetoday.com/resource-centers/advances-in-dermatology/individuals-color-have-poor-melanoma-survival/1375*.

Student Handout

TASK 16. WHAT IS MELANOMA?

Introduction

So far, you have learned that Freja died from melanoma, a form of skin cancer. You need to find more information about cancer to understand how Freja may have gotten it and how it affected her. In this task, you will explore the different types of skin cancer and determine potential risks for cancer of virtual patients.

Initial Ideas

Before beginning this task, take a few minutes to think about skin cancer. What is skin cancer?

What causes it?

What are the risk factors for getting skin cancer?

How can you prevent it? Who can get skin cancer?

Your Task

Your task is to use the internet to discover information about skin cancer. As you search and read through the information, answer the questions below.

Healthy Skin

What are the three main layers of skin?

Which layer helps protect against sunlight and radiation?

Causes

How would you define skin cancer? What are the three types of skin cancer?

Illustrate the long-term effects of ultraviolet (UV) radiation on a piece of DNA in the boxes below.

Long-Term Effects of UV Radiation on DNA

DNA before exposure to UV radiation:	DNA after exposure to UV radiation:

What effect does this damage to DNA have on skin cells?

Tumors

Compare and contrast benign and malignant tumors.

Risk Factors

Briefly explain how each of these risk factors is associated with skin cancer, and answer the question about Freja's risk factors.

Age	Complexion
Environment	Genetics
Actinic keratosis	Which of these risk factors did Freja have?

Melanoma

Summarize the characteristics, risk factors, and symptoms of *melanoma*, the type of cancer Freja had.

Characteristics	Risk factors	Symptoms

Prevention

Briefly explain how each of the following can help prevent skin cancer.

Sunglasses	Sunscreen
Hat	Limiting exposure

The Biology of Skin Color

Watch the HHMI BioInteractive video "The Biology of Skin Color" at *www.biointeractive.org/classroom-resources/biology-skin-color*. Then answer the following questions.

1. What causes differences in skin color?

2. How is skin color a trade-off between protection from UV and the need for some UV absorption for the production of vitamin D?

3. Based on what you know about the causes of skin cancer, how does skin color affect your chances of getting skin cancer?

Some Useful Ideas From My Teacher

You can keep track of useful ideas from your teacher in the space below.

What We Figured Out

Now that you have completed this task, take a few minutes to fill out the Summary Table below with the other students in your group. This table will help you keep track of what you figured out during the task. You will then have an opportunity to share your ideas as we fill out our class Summary Table.

Summary Table

What we learned	How we know

How it helps us explain the phenomenon

Teacher Notes

Task 17. Freja's Factors

Purpose

The purpose of this task is to give students the opportunity to determine which risk factors likely contributed to Freja's development of melanoma. In the previous task, students learned about the risk factors for melanoma. This task provides students with data from research studies to analyze and compare the risk factors for the development of melanoma.

Important Life Science Content

Analysis of the data from various sources provides insight into the factors that may have contributed to Freja's development of skin cancer. A meta-analysis of several studies examining the risk of melanoma associated with indoor tanning equipment found a positive association between the use of tanning equipment and the risk of melanoma. It was reported that Freja had used tanning beds at some point during her life, so this may have had some effect on the development of her melanoma. Analysis of data from around the world shows that the geographic locations with the highest incidence of melanoma are New Zealand and Australia; Europe, including England, where Freja lived; and North America. Age is also known to be a risk factor, and the data show a positive correlation between increasing age and risk of melanoma. Individuals in the age 15–39 category, to which Freja belonged, had the lowest risk of developing skin cancer, so this was not a factor for her. The American Cancer Society (2021) states that melanoma is "more than 20 times more common in whites than African Americans." This indicates that Freja's fair skin may have also contributed to the development of her cancer. Finally, the results of an Australian study show that the use of sunscreen decreases the risk of developing melanoma. The fact that Freja did not use sunscreen all the time was likely a contributing factor to her developing melanoma.

Scientific Ideas That Are Important to Think About During This Task

- There are several risk factors that contribute to melanoma.

Timeline

Approximately one class period.

Materials and Preparation

The items needed for this investigation are listed in Table 3.5. In preparation for this task, we recommend that you familiarize yourself with the sources of the data on the Student Handout.

Table 3.5. Required Materials for Task 17

Item	Quantity
Student Handout	1 per student

Safety Precautions

This task does not require any specific safety precautions.

Procedure

This lesson plan is only a suggestion. It is included here to illustrate how you can facilitate student thinking during this task. We encourage you to modify this lesson plan by asking different questions, using different examples, and providing different scaffolds as appropriate to better meet the needs of students in your class.

Putting Ideas on the Table (50 minutes)

1. Introduce the task with the PowerPoint template we have provided, which can be downloaded from the book's Extras page at *www.nsta.org/mbi-biology*.

2. Each group should work on interpreting the graphs on the Student Handout.

3. During the task, be sure to walk around to each group using appropriate Back Pocket Questions below to press students to think more deeply about the risk factors for melanoma.

4. Once the groups have completed the task, conduct a whole-class discussion to help make sense of the data in relation to Freja's story.

Adding Information to the Summary Table (10 minutes)

1. Give students 5 minutes to decide what to add to the Summary Table at the end of the Student Handout.

2. Have one student from each group share what the group figured out, how they know (their evidence for what they figured out), and how this information will help them explain the anchoring phenomenon.

3. Once each group has shared, ask the entire class to decide what should be added to each column of the class Summary Table for Task 17. Help students reach consensus about what to add to the Summary Table. Only add an idea to the Summary Table if everyone in the class agrees with that idea.

Back Pocket Questions

As students work in groups, it is important to engage with each group to help press and extend students' thinking around the ideas at play in this task. Following are some example questions you might ask:

1. Helping students get started: How would you describe a risk factor? What risk factors are you surprised by?

2. Pressing further: What patterns are you seeing in the data? Which factors are the most relevant to Freja's story?

3. Following up: What makes you think that? Can you say more?

Filling Out the Summary Table

Table 3.6 includes examples of the responses students may come up with when they fill out the Summary Table. This is provided here only as a behind-the-scenes roadmap and is not meant to be shared with students.

Table 3.6. Example Summary Table for Task 17

What we learned	How it helps us explain the phenomenon
There are several risk factors that contribute to melanoma. These include increased age, use of tanning beds, geographic location (Europe, Australia, North America), light skin color, and lack of sunscreen use.	Freja had several risk factors that likely contributed to her melanoma. She had light skin and lived in England. She had used tanning beds and often did not use sunscreen while at home. Her age was not a risk factor that contributed to her melanoma.

Possible Extension

- Use UV beads, UV paper, or bananas to test the effectiveness of sunblock.

Reference

American Cancer Society. 2021. Key statistics for melanoma skin cancer. Last updated January 12, 2021. *www.cancer.org/cancer/melanoma-skin-cancer/about/key-statistics.html.*

Student Handout

TASK 17. FREJA'S FACTORS

Introduction

In the previous task, you found out about the possible risk factors that contribute to developing melanoma. In this task, you will analyze data to determine what assumptions in your model are supported by evidence.

Initial Ideas

Before beginning this task, take a few minutes to think about the risk factors for melanoma. What risk factors for melanoma do you think Freja had?

Your Task

Your task is to analyze the data provided in Figures SH17.1–SH17.5 to determine the risk factors associated with the development of melanoma.

Figure SH17.1. Relative Risk for Cutaneous Melanoma Associated With First Use of Indoor Tanning Equipment at Age <35 Years: Estimates of 7 Studies and Summary Estimate

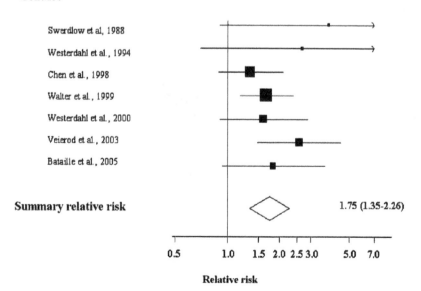

Analysis of Figure SH17.1: What information does this figure tell you about the risk of skin cancer as related to the use of tanning equipment?

Figure SH17.2. Worldwide Melanoma Age-Standardized Annual Incidence Rate by Geography

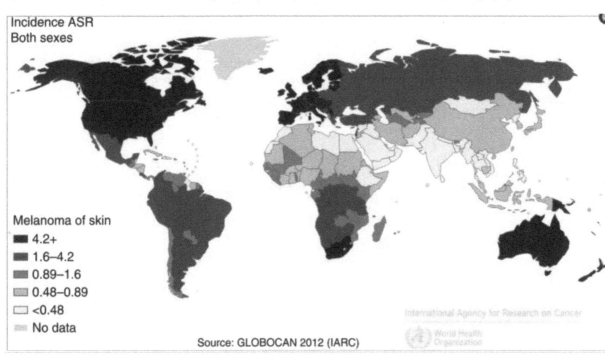

Note: Age-standardized rate (ASR) by world is expressed per 100,000 persons.

Analysis of Figure SH17.2: What information does this figure tell you about the risk of skin cancer as related to geography?

Figure SH17.3. Worldwide Age-Standardized Incidence of Melanoma by Age

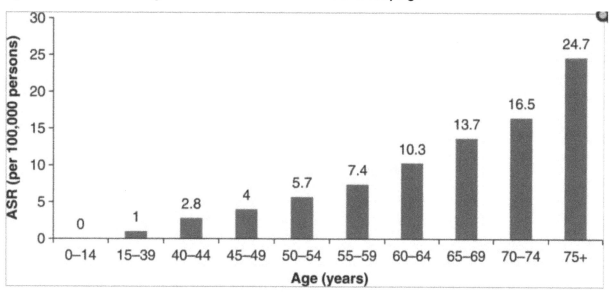

Note: Age-standardized rate (ASR) by world is expressed per 100,000 persons.

Analysis of Figure SH17.3: What information does this figure tell you about the risk of skin cancer as related to age?

Figure SH17.4. Key Statistics on the Risk of Getting Melanoma From the American Cancer Society (2021)

Melanoma is more than 20 times more common in whites than in African Americans. Overall, the lifetime risk of getting melanoma is about 2.6% (1 in 38) for whites, 0.1% (1 in 1,000) for Blacks, and 0.6% (1 in 167) for Hispanics.

Analysis of Figure SH17.4: What does the information from the American Cancer Society in this figure tell you about the risk of skin cancer as related to skin color?

Figure SH17.5. Excerpt From Journal Article "Sunscreen Use and Melanoma Risk Among Young Australian Adults"

Risk of melanoma was less with higher use of sunscreen in childhood (OR for highest vs lowest tertiles, 0.60; 95% CI, 0.42–0.87; $P = .02$ for trend) and across the lifetime (OR, 0.65; 95% CI, 0.45–0.93; $P = .07$ for trend). Subgroup analyses suggested that the protective association of sunscreen with melanoma was stronger for people reporting blistering sunburn, receiving a diagnosis of melanoma at a younger age, or having some or many nevi. Total lifetime sun exposure was unrelated to melanoma risk (OR for highest vs lowest tertile, 0.97; 95% CI, 0.66–1.43; $P = .94$ for trend). By contrast, total sun exposure inversely weighted by sunscreen use (as a measure of sun exposure unprotected by sunscreen) was significantly associated with melanoma risk (OR, 1.80; 95% CI, 1.22–2.65; $P = .007$ for trend) and appeared stronger for people having lighter pigmentation or some or many nevi or using sunscreen to stay longer in the sun.

Analysis of Figure SH17.5: What does the information from the journal article in this figure tell you about the risk of skin cancer as related to sunscreen use?

Some Useful Ideas From My Teacher

You can keep track of useful ideas from your teacher in the space below.

What We Figured Out

Now that you have completed this task, take a few minutes to fill out the Summary Table below with the other students in your group. This table will help you keep track of what you figured out during the task. You will then have an opportunity to share your ideas as we fill out our class Summary Table.

Summary Table

What we learned	How we know

How it helps us explain the phenomenon

Reference

American Cancer Society. 2021. Key statistics for melanoma skin cancer. Last updated January 12, 2021. *www.cancer.org/cancer/melanoma-skin-cancer/about/key-statistics.html.*

Teacher Notes

Task 18. Control of the Cell Cycle Online Simulation

Purpose

In the previous tasks, students learned about skin cancer. They discovered that cancerous cells are different from normal cells. This task helps students make sense of why cancerous cells are different. Students learn about the cell cycle and how it is regulated in normal cells, as well as what changes occur to the cell cycle in cancerous cells. They also learn that mutations in genes that control the cell cycle are responsible for the development of cancer. In the next task, students will learn about the structure of genes.

Important Life Science Content

During their lifetime, cells progress through the cell cycle, in which they grow, develop, and prepare for cellular division at the end of mitosis. The first phase of the cell cycle is Gap 1 (G1), when the cell grows, copies its organelles, and makes molecular building blocks to be used later in the cycle. Next is the synthesis phase (S), in which the cell copies its DNA and centrosomes. Following S is Gap 2 (G2), when the cell grows, makes proteins and organelles, and prepares for mitosis. The last stage of the cell cycle is mitosis (M), during which the cell divides its chromosomes and cytoplasm and goes through cytokinesis to produce two cells.

There are several checkpoints within the cell cycle that are used to regulate the cell's progress through the cycle. If cells meet the criteria at each checkpoint, they will proceed to the next stage. If they do not pass the checkpoint, the cell remains in that stage until it meets the criteria or it dies. The first checkpoint is found at the end of G1. The G1 checkpoint looks for several things: Is the cell large enough? Does the cell have enough energy and nutrients? Is the cell receiving positive signals from its environment? Is there any damage to the DNA? If the G1 checkpoint is cleared, the cell proceeds to the S phase of the cycle. Next, the G2 checkpoint prevents cells with damaged DNA from entering mitosis. It checks to see if the DNA completely copied during the S phase or if it was damaged. Cells that pass the G2 checkpoint move on to mitosis, where the final checkpoint in the cell cycle makes sure all the chromosomes are attached to the spindle microtubules. Once this checkpoint is cleared, the cell progresses through mitosis and cytokinesis, resulting in the formation of two daughter cells.

Cells that are cancerous are different from normal cells because their cell cycle is not regulated. The molecules cyclin and cyclin-dependent kinase (CDK) are important in regulating the cell cycle. Loss of control of the cell cycle can be related to these or other regulatory proteins being ineffective. Rather than stopping at the checkpoints, the cells progress through the cell cycle continuously. This results in constant progression toward cell division and the overproduction of cells, which leads to tumor formation.

Scientific Ideas That Are Important to Think About During This Task

- Normal cells go through several checkpoints in the cell cycle that regulate the timing of cell division.

Timeline

Approximately one class period.

Materials and Preparation

The items needed for this investigation are listed in Table 3.7. In preparation for this task, we recommend that you familiarize yourself with the online simulation and make sure it can be accessed on student devices.

Table 3.7. Required Materials for Task 18

Item	Quantity
Computer or tablet with internet connection	1 per group
Student Handout	1 per student

Safety Precautions

This task does not require any specific safety precautions.

Procedure

This lesson plan is only a suggestion. It is included here to illustrate how you can facilitate student thinking during this task. We encourage you to modify this lesson plan by asking different questions, using different examples, and providing different scaffolds as appropriate to better meet the needs of students in your class.

Putting Ideas on the Table (50 minutes)

1. Introduce the task with the PowerPoint template we have provided, which can be downloaded from the book's Extras page at *www.nsta.org/mbi-biology*.

2. Groups should access "The Eukaryotic Cell Cycle and Cancer" simulation on the HHMI BioInteractive website at *www.biointeractive.org/classroom-resources/eukaryotic-cell-cycle-and-cancer*.

3. Each group should work through the online simulation and answer the appropriate questions on the Student Handout.

4. During the task, be sure to walk around to each group using appropriate Back Pocket Questions (see next page) to press students to think more deeply about the cell cycle and its relationship to cancer.

5. Once the groups have completed the task, conduct a whole-class discussion to help students make sense of the data in relation to Freja's story.

Adding Information to the Summary Table (10 minutes)

1. Give students 5 minutes to decide what to add to the Summary Table at the end of the Student Handout.

2. Have one student from each group share what the group figured out, how they know (their evidence for what they figured out), and how this information will help them explain the anchoring phenomenon.

3. Once each group has shared, ask the entire class to decide what should be added to each column of the class Summary Table for Task 18. Help students reach consensus about what to add to the Summary Table. Only add an idea to the Summary Table if everyone in the class agrees with that idea.

Back Pocket Questions

As students work in groups, it is important to engage with each group to help press and extend students' thinking around the ideas at play in this task. Following are some example questions you might ask:

1. Helping students get started: How is the cell cycle of normal cells different from that of cancerous cells?

2. Pressing further: How is this connected to what was happening in Freja's cells?

3. Following up: What makes you think that? Can you say more?

Filling Out the Summary Table

Table 3.8 includes examples of the responses students may come up with when they fill out the Summary Table. This is provided here only as a behind-the-scenes roadmap and is not meant to be shared with students.

Table 3.8. Example Summary Table for Task 18

What we learned	How it helps us explain the phenomenon
Normal cells go through several checkpoints in the cell cycle that regulate the timing of cell division. Cancerous cells are not regulated by the checkpoints and continuously move through the cell cycle without stopping. This leads to the overproduction of cells that eventually form a tumor.	Cancerous cells are not regulated by the cell cycle checkpoints and divide continuously. Freja's cancerous cells were dividing out of control and formed tumors.

Hints for Implementing This Task

- The first six questions are from the introductory section on the website, which is divided into four parts. Tell students that once they enter the game, they will not be able to go back and review this section unless they refresh their screen and start over.

- If students choose the wrong options in the simulation, it will eventually send them back out to learn more about the cell cycle before continuing.

Possible Extensions

- Students may do the "Cancer and the Cell Cycle" coloring task at *www.teacherspayteachers. com/Product/Cancer-and-the-Cell-Cycle-1260070*.

- Have students watch "Amoeba Sisters Unlectured Series: Cell Cycle" at *www.teacherspay-teachers.com/Product/Amoeba-Sisters-Unlectured-Series-CELL-CYCLE-3920355*.

Student Handout

TASK 18. CONTROL OF THE CELL CYCLE ONLINE SIMULATION

Introduction

During a cell's lifetime, it progresses through the stages of the cell cycle, which directs cell growth, development, and division. Checkpoints along the way are used to regulate the progression through the stages (G1, S, G2, and M). Previously, you learned that cancerous cells divide rapidly. In this task, you will learn how this is connected to the cell cycle.

Initial Ideas

Before beginning this task, take a few minutes to think about cancer in relation to the cell cycle. How does the cell regulate its progression through the cell cycle?

How does the cell cycle in cancerous cells compare with the cycle for normal cells?

Your Task

Your task is to explore the cell cycle in normal cells and cancerous cells. Go to the following website and click on Launch Interactive: *www.biointeractive.org/classroom-resources/eukaryotic-cell-cycle-and-cancer*. Answer the following questions as you explore. Be sure to read through EVERYTHING in the online simulation! Don't forget to look at the Background and Key Concepts tabs on the right for additional information.

1. Identify two reasons that cells divide.

2. How many cells does an average human body contain?

3. When do cells generally stop dividing and exit the cell cycle?

4. How does a cell know when to start dividing?

5. What key molecules control and coordinate cell division?

On the cell cycle diagram, choose the G1 Checkpoint. Continue clockwise through the other phases as you answer the questions below.

1. What occurs in the first gap phase (G1)?

2. At the end of the G1 phase, what things must be checked before moving on? If they are not checked, what happens?

3. What is the next step in the process if the cell is ready to divide? What is the product of this phase, and what does the cell check to make sure it can move to the next phase?

4. What happens during the second gap phase (G2)?

TASK 18. CONTROL OF THE CELL CYCLE ONLINE SIMULATION

5. Explain what is checked in checkpoint 2 after the G2 phase.

6. When the cell enters mitosis (M phase), what happens to the chromosomes?

7. What is checked at the end of mitosis (M phase)?

8. What happens when checkpoint 3 is cleared?

On the cell cycle diagram, choose Cell Cycle Regulators and Cancer (in the middle of the diagram) to answer the questions below. (Notice the Cancer Overview tab.)

1. What's the difference between a stimulating protein and an inhibitory protein?

2. Explain what CDKs and cyclins do for the cell.

3. Explain what causes cancer and how tumors form.

4. If you were asked to explain to a 5-year-old what causes cancer, what would you say?

Some Useful Ideas From My Teacher

You can keep track of useful ideas from your teacher in the space below.

What We Figured Out

Now that you have completed this task, take a few minutes to fill out the Summary Table below with the other students in your group. This table will help you keep track of what you figured out during the task. You will then have an opportunity to share your ideas as we fill out our class Summary Table.

Summary Table

What we learned	How we know

How it helps us explain the phenomenon

Midunit Model Revision for Unit 3

Purpose

Now that your class is roughly halfway through the unit, it is important for students to go back to the initial models constructed on the first day and revise them based on what they have learned so far. Students may choose between two different strategies for model revisions. After negotiating what should be added, revised, or removed from their models, the groups may do either of the following:

- Redraw their models.
- Use sticky notes to keep track of their revisions for the final model revision near the end of the unit.

They should make the choice between these options based on the amount of time available for the model revision process. It is important, however, that students make decisions about revisions based on what fits with the evidence from the tasks or is consistent with the scientific ideas at play. We have included a model revision PowerPoint template for your use, which can be downloaded from the book's Extras page at *www.nsta.org/mbi-biology*, and Figure 3.7 contains resources to help with modeling.

Figure 3.7. Resources to Help With Modeling

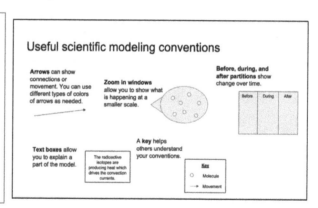

Although it does take time, we recommend a share-out session after the model revision. This allows for new ideas to be shared across the whole class, as well as discussion about what should and should not be included in the models. Have the members of each group stand up and describe how and why they changed their model. If time is limited, you could ask each group to focus on just one change. Provide time for questions from the class, and be sure to ask clarifying questions while comparing and contrasting the ideas presented across the groups.

Teacher Notes

Task 19. DNA Structure Pop Bead Model

Purpose

The purpose of this task is to introduce students to the structure of the DNA molecule. Previously, students learned that DNA in skin cells can be damaged by UV radiation from the Sun. They also learned that changes in genes may alter the normal functioning of the cell cycle. This task is designed to help students begin making sense of what genes really are. Following this lesson, students will dive deeper into the molecular biology of protein synthesis. To understand that concept, they must have a strong understanding of the structure of DNA.

Important Life Science Content

Deoxyribonucleic acid, or DNA, is the genetic material found in all living things. The molecule consists of two chains of nucleotides that are connected in the middle and twisted around each other in a double helix shape. The monomer subunit that makes up the chain of DNA is the nucleotide. Nucleotides have three components: a deoxyribose sugar, a phosphate group, and a nitrogen base. The backbone of the structure consists of alternating deoxyribose and phosphate molecules that are bonded together through phosphodiester linkages at the 5' and 3' carbons of the sugar molecule. The nitrogen bases of the nucleotides stick out into the center of the double helix and connect the two strands of DNA together through hydrogen bonds.

Key to the process of DNA replication and transcription is the fact that the nitrogen bases pair up in a complementary fashion. Adenine always pairs with thymine, and cytosine always pairs with guanine. Because of the complementary pairing that occurs, if one side of the DNA molecule is known, the other side can be constructed. Although the basic structure of DNA is the same in all organisms, the sequence of nitrogen bases varies. This provides for the diversity of living things.

Scientific Ideas That Are Important to Think About During This Task

- All cells contain genetic information in the form of DNA molecules.

Timeline

Approximately one class period.

Materials and Preparation

The items needed for this investigation are listed in Table 3.9 (p. 282). You can purchase the pop beads from a science supply company such as Flinn Scientific (*www.flinnsci.com/products/biology/genetics--dna/pop-beads*).

Table 3.9. Required Materials for Task 19

Item	Quantity
Safety glasses or goggles	1 per student
White pop beads	22 per group
Red pop beads	22 per group
Orange pop beads	5–10 per group
Green pop beads	5–10 per group
Blue pop beads	5–10 per group
Yellow pop beads	5–10 per group
Clear pop beads	1 per group
Student Handout	1 per student

To save time, prepare the materials in advance for each group: 22 deoxyribose pop beads (white), 22 phosphate pop beads (red), 5–10 of each of the nitrogen base pop beads (orange, green, blue, and yellow), and 11 connectors (clear).

Safety Precautions

Remind students to follow all normal safety rules. In addition, tell students to take the following safety precautions:

- Wear safety glasses or goggles during setup, hands-on activity, and cleanup.
- Immediately pick up any items dropped on the floor so they do not become a slip or fall hazard.
- Wash hands with soap and water when the activity is completed.

Procedure

This lesson plan is only a suggestion. It is included here to illustrate how you can facilitate student thinking during this task. We encourage you to modify this lesson plan by asking different questions, using different examples, and providing different scaffolds as appropriate to better meet the needs of students in your class.

Putting Ideas on the Table (50 minutes)

1. Introduce the task with the PowerPoint template we have provided, which can be downloaded from the book's Extras page at *www.nsta.org/mbi-biology*.

2. Each group should work through the pop bead modeling and answer the appropriate questions on the Student Handout.

3. During the task, be sure to walk around to each group, using appropriate Back Pocket Questions (below) to press students to think more deeply about the cell cycle and its relationship to cancer.

4. Once the groups have completed the task, conduct a whole-class discussion to help make sense of the data in relation to Freja's story.

Adding Information to the Summary Table (10 minutes)

1. Give students 5 minutes to decide what to add to the Summary Table at the end of the Student Handout.

2. Have one student from each group share what the group figured out, how they know (their evidence for what they figured out), and how this information will help them explain the anchoring phenomenon.

3. Once each group has shared, ask the entire class to decide what should be added to each column of the class Summary Table for Task 19. Help students reach consensus about what to add to the Summary Table. Only add an idea to the Summary Table if everyone in the class agrees with that idea.

Back Pocket Questions

As students work in groups, it is important to engage with each group to help press and extend students' thinking around the ideas at play in this task. Following are some example questions you might ask:

1. Helping students get started: How do the base pairing rules help you construct the DNA molecule?

2. Pressing further: What would happen if you used a different order of nitrogen bases in your DNA strand?

3. Following up: What makes you think that? Can you say more?

Filling Out the Summary Table

Table 3.10 (p. 284) includes examples of the responses students may come up with when they fill out the Summary Table. This is provided here only as a behind-the-scenes roadmap and is not meant to be shared with students.

Table 3.10. Example Summary Table for Task 19

What we learned	How it helps us explain the phenomenon
DNA is a molecule consisting of two chains. Attached to the backbone of each chain are nitrogen bases (A, T, C, G). Each nitrogen base is bonded together with a complementary base (A-T and C-G). The structure of DNA allows it to be copied as the complementary bases pair up.	This helps us understand the directions in the DNA of cells and what the mutations might affect. The genes involved in Freja's cancer are coded by the DNA nucleotide sequences.

Hints for Implementing This Task

- Alternative DNA model kits or paper cutouts can be used in place of the pop bead models. Printables can be purchased at *www.teacherspayteachers.com/Product/ Create-Paper-DNA-Structure-Model-Task-Simple-Instructions-NGSS-LS3-1-3654937*.

Possible Extensions

- If you would like students to explore the structure of DNA using a more inquiry-based approach, consider using Lab 18 in *Argument-Driven Inquiry in Biology: Lab Investigations for Grades 9–12* at *www.nsta.org/store/product_detail.aspx?id=10.2505/9781938946202*.

- Have students model DNA replication by connecting their models of DNA together to make one long strand. You can also use these connected models to demonstrate that genes are located on sections of DNA.

Student Handout

TASK 19. DNA STRUCTURE POP BEAD MODEL

Introduction

Deoxyribonucleic acid, or DNA, is the hereditary molecule. It is the genetic material of living organisms and is located in the chromosomes of each cell. DNA is made of individual units called nucleotides, which are the building blocks of DNA. Each nucleotide is made of a phosphate group, a sugar (deoxyribose), and a nitrogen base. The phosphate and sugar form the backbone (sides) of the molecule. Each rung (the steps of the DNA ladder) contains a pair of bases held together by hydrogen bonds. There are four bases: thymine (T), adenine (A), guanine (G), and cytosine (C). T and A always pair up, and G and C always pair up.

Initial Ideas

Before beginning this task, take a few minutes to think about the DNA molecule. What is the structure of DNA?

Your Task

Your task is to work as a group to model the structure of DNA using beads to represent the different parts of the DNA molecule.

Safety Precautions

Follow all normal lab safety rules. In addition, be sure to take the following safety precautions:

- Wear safety glasses or goggles during setup, hands-on activity, and cleanup.
- Immediately pick up any items dropped on the floor so they do not become a slip or fall hazard.
- Wash hands with soap and water when the activity is completed.

Activity

1. Wearing safety glasses or goggles, begin by creating a nucleotide consisting of one phosphate, one deoxyribose sugar, and one nitrogen base (adenine, thymine, guanine, or cytosine). Connect the pieces as listed in Table SH19.1 (p. 286).

Table SH19.1. Components Represented by Pop Beads

Component	Pop Bead Color
Deoxyribose (sugar)	White
Phosphate	Red
Adenine (nitrogen base)	Yellow
Thymine (nitrogen base)	Green
Guanine (nitrogen base)	Orange
Cytosine (nitrogen base)	Blue
Hydrogen bond	Clear connectors

- Repeat step 1 to make 10 more nucleotides, randomly picking a nitrogen base each time.

2. Join the nucleotides together so that the phosphates and sugars alternate (for example, red-white-red-white). This will be your original DNA strand. At this point, the order of the bases does not matter.

3. Once you have created the original DNA strand, complete the first column of the DNA Molecule Data Table on the next page. Write the letter of the base, followed by the color of the bead, in the order in which they appear in your model.

4. Next, determine the base sequence for the complementary DNA strand. Use Table SH19.1 to help. Remember that **T**hymine pairs with **A**denine, and **G**uanine pairs with **C**ytosine. In the DNA Molecule Data Table, indicate the letter of the complementary base and the color of the corresponding pop bead.

5. One at a time, create the complementary nucleotides that you will attach to the original strand. Once you have created a nucleotide, connect it to the complementary nucleotide on the original strand using the clear plastic connectors, which represent the hydrogen bonds that keep the base pairs together.

6. Holding the model from the top, gently twist the DNA ladder to the right. You should see that the DNA looks like a spiral staircase. The model now represents the double helix shape of DNA.

7. Have your *teacher check* that you have completed the DNA model correctly and initial your worksheet.

8. In the last column in the DNA Molecule Data Table, draw your DNA molecule untwisted. Label and color-code your sketch.

DNA Pop Bead Model Data Collection

Model is correct: _____

(Teacher Initials)

DNA Molecule Data Table

Original strand (Strand #1)	Complementary strand (Strand #2)	Your Drawing
Nitrogen base	Complementary base	Draw your model in the space below.
Ex: C (blue)	Ex: G (orange)	Label a phosphate, a sugar, and one of each nitrogen base. Don't forget to color-code your sketch!

Data Analysis

1. Circle and label a single nucleotide on your drawing of your model.

2. Which three molecules make up a nucleotide?

3. Which molecules make up the backbone of the DNA molecule?

4. What type of bond keeps the bases paired together?

5. Which base always pairs with adenine?

6. Which base always pairs with cytosine?

7. If you had a DNA strand with a sequence of **T A T T G G C C A C G T,** what would the bases on the complementary strand be?

8. Compare your model with another group's model. What similarities and differences can you observe?

9. All living things contain DNA, yet they are vastly different from each other. What have you learned in this task that helps explain why there are such differences among living things?

Some Useful Ideas From My Teacher

You can keep track of useful ideas from your teacher in the space below.

What We Figured Out

Now that you have completed this task, take a few minutes to fill out the Summary Table below with the other students in your group. This table will help you keep track of what you figured out during the task. You will then have an opportunity to share your ideas as we fill out our class Summary Table.

Summary Table

What we learned	How we know

How it helps us explain the phenomenon

Teacher Notes

Task 20. What Are Proteins?

Purpose

The purpose of this task is to have students explore how proteins are made using the template provided on a gene. They also discover that proteins have a wide variety of functions. Proteins are the result of a particular sequence of DNA contained on genes in cells. From the previous task, students already understand that DNA is made up of four different nucleotides and that the sequence of nucleotides varies in DNA molecules. In this task, they explore how genes and proteins are connected. They also learn that cancer is caused by tumor suppressor genes and oncogenes that affect a variety of cell functions involved in the cell cycle. In the next task, students will explore the mechanics of this process in greater depth.

Important Life Science Content

Proteins are biomolecules essential to the structure and function of living things. "The structure of DNA determines the structure of proteins, which carry out the essential functions of life through systems of specialized cells" (HS-LS1-1). The sequence of DNA nucleotides affects the sequence of amino acids in a protein. If the sequence of amino acids is different, then it results in a different protein with a different function.

Scientific Ideas That Are Important to Think About During This Task

- The sequence of nucleotides determines the sequence of amino acids in a protein.
- Each protein serves a specific function within an organism.

Timeline

Approximately one class period.

Materials and Preparation

The items needed for this investigation are listed in Table 3.11. Make copies of the protein profiles, codon wheel, and worksheet found at *www.yourgenome.org/activities/function-finders*.

Table 3.11. Required Materials for Task 20

Item	Quantity
Protein profiles	1 per group
Codon wheel	1 per student
Worksheet	1 per student
Student Handout	1 per student

Safety Precautions

This task does not require any specific safety precautions.

Procedure

This lesson plan is only a suggestion. It is included here to illustrate how you can facilitate student thinking during this task. We encourage you to modify this lesson plan by asking different questions, using different examples, and providing different scaffolds as appropriate to better meet the needs of students in your class.

Putting Ideas on the Table (50 minutes)

1. Introduce the task with the PowerPoint template we have provided, which can be downloaded from the book's Extras page at *www.nsta.org/mbi-biology*.

2. Each group should work through the task to translate the DNA sequences to specific proteins.

3. During the task, be sure to walk around to each group using appropriate Back Pocket Questions (below) to press students to think more deeply about the cell cycle and its relationship to cancer.

4. Once the groups have completed the task, conduct a whole-class discussion to help make sense of the data in relation to Freja's story.

Adding Information to the Summary Table (10 minutes)

1. Give students 5 minutes to decide what to add to the Summary Table at the end of the Student Handout.

2. Have one student from each group share what the group figured out, how they know (their evidence for what they figured out), and how this information will help them explain the anchoring phenomenon.

3. Once each group has shared, ask the entire class to decide what should be added to each column of the class Summary Table for Task 20. Help students reach consensus about what to add to the Summary Table. Only add an idea to the Summary Table if everyone in the class agrees with that idea.

Back Pocket Questions

As students work in groups, it is important to engage with each group to help press and extend students' thinking around the ideas at play in this task. Following are some example questions you might ask:

1. Helping students get started: How are proteins and DNA structure related?

2. Pressing further: If the DNA in Freja's cells were damaged, what effect do you think that might have on the proteins in Freja's cells?

3. Following up: What makes you think that? Can you say more?

Filling Out the Summary Table

Table 3.12 includes examples of the responses students may come up with when they fill out the Summary Table. This is provided here only as a behind-the-scenes roadmap and is not meant to be shared with students.

Table 3.12. Example Summary Table for Task 20

What we learned	How it helps us explain the phenomenon
The sequence of nucleotides determines the sequence of amino acids in a protein. Each protein serves a specific function within an organism.	The proteins in Freja's skin cells were made using the directions in her genes. Specific proteins, such as p53, may be important in producing cells that divide properly.

Possible Extensions

- Look at other genes and proteins involved in melanoma patients, such as BRAF, NRAS, MLL3, and TP53, by conducting research at the following websites:
  - *www.ncbi.nlm.nih.gov*
  - *www.uniprot.org*
  - *https://blast.ncbi.nlm.nih.gov/Blast.cgi?PROGRAM=blastp&PAGE_TYPE= BlastSearch&LINK_LOC=blasthome*
  - *https://learn.genetics.utah.edu/content/basics/transcribe.*

Student Handout

TASK 20. WHAT ARE PROTEINS?

Introduction

DNA contains the genetic information needed for your body to function. Sections of DNA that code for proteins are called genes. Humans have between 20,000 and 25,000 genes. The sequence of nucleotides on each gene codes for a unique protein. Every protein has a particular function. In the last task, you constructed a molecule of DNA. In this task, you will discover how that structure serves as the code for proteins.

Initial Ideas

Before beginning this task, take a few minutes to think about DNA and proteins. How does the structure of DNA determine the function of a protein?

What are some functions proteins can have?

Your Task

Your task is to look at the first few nucleotides of actual genes found in a variety of organisms and decode each gene to show the protein it produces, then determine the function of the protein. Your teacher will provide you with copies of the protein profiles, codon wheel, and worksheet that you will need to complete this task.

Some Useful Ideas From My Teacher

You can keep track of useful ideas from your teacher in the space below.

What We Figured Out

Now that you have completed this task, take a few minutes to fill out the Summary Table below with the other students in your group. This table will help you keep track of what you figured out during the task. You will then have an opportunity to share your ideas as we fill out our class Summary Table.

Summary Table

What we learned	How we know

How it helps us explain the phenomenon

Teacher Notes

Task 21. Protein Synthesis Simulation

Purpose

The purpose of this task is to help students learn how cells make proteins by using an online simulation. They have already learned that proteins have a variety of functions and originate from genes that have a specific nucleotide sequence. Students connect this information to the driving question through a discussion of how tumor suppressor genes and oncogenes create proteins that allow the cell cycle checkpoints to work.

Important Life Science Content

Before amino acids are formed, messenger RNA (mRNA) is produced from the original DNA strand. DNA cannot leave the nucleus, so mRNA delivers the message from the DNA to the ribosomes in the cytoplasm of the cell. Here, the amino acids are delivered to the ribosome, one by one, by transfer RNA (tRNA) molecules, and an amino acid chain is created. After the amino acid chain is produced in the ribosome, the protein is released and folds into a unique shape.

Scientific Ideas That Are Important to Think About During This Task

- DNA is transcribed into mRNA, which is translated into amino acids to create different proteins.
- The sequence of DNA nucleotides on a gene affects the resulting protein.

Timeline

Approximately one class period.

Materials and Preparation

The items needed for this investigation are listed in Table 3.13. In preparation for this task, we recommend that you familiarize yourself with the online simulation and make sure it can be accessed on student devices.

Table 3.13. Required Materials for Task 21

Item	Quantity
Computer or tablet with internet connection	1 per group
Student Handout	1 per student

Safety Precautions

This task does not require any specific safety precautions.

Procedure

This lesson plan is only a suggestion. It is included here to illustrate how you can facilitate student thinking during this task. We encourage you to modify this lesson plan by asking different questions, using different examples, and providing different scaffolds as appropriate to better meet the needs of students in your class.

Putting Ideas on the Table (50 minutes)

1. Introduce the task with the PowerPoint template we have provided, which can be downloaded from the book's Extras page at *www.nsta.org/mbi-biology*.

2. Ask students to work together in groups on the Concord Consortium's simulation at *https://authoring.concord.org/activities/22/pages/113/c45b130e-5e22-40d8-bcdc-74210b536a3d*, discussing appropriate responses for the questions on the Student Handout as they work through the online presentation.

3. Rotate among the groups, asking appropriate Back Pocket Questions (below) to help students make sense of the content.

4. At the end of the task, conduct a whole-class discussion to aid student understanding of skin cancer.

Adding Information to the Summary Table (10 minutes)

1. Give students 5 minutes to decide what to add to the Summary Table at the end of the Student Handout.

2. Have one student from each group share what the group figured out, how they know (their evidence for what they figured out), and how this information will help them explain the anchoring phenomenon.

3. Once each group has shared, ask the entire class to decide what should be added to each column of the class Summary Table for Task 21. Help students reach consensus about what to add to the Summary Table. Only add an idea to the Summary Table if everyone in the class agrees with that idea.

Back Pocket Questions

As students work in groups, it is important to engage with each group to help press and extend students' thinking around the ideas at play in this task. Following are some example questions you might ask:

1. Helping students get started: What is the sequence of steps in protein synthesis? What happened before what you are doing right now? What do you think will happen next?

2. Pressing further: How does this information link to what you already know about the genes involved in causing melanoma?

3. Following up: What makes you think that? Can you say more?

Filling Out the Summary Table

Table 3.14 includes examples of the responses students may come up with when they fill out the Summary Table. This is provided here only as a behind-the-scenes roadmap and is not meant to be shared with students.

Table 3.14. Example Summary Table for Task 21

What we learned	How it helps us explain the phenomenon
DNA is transcribed into mRNA, which is translated into amino acids to create different proteins. The sequence of DNA nucleotides on a gene affects the resulting protein.	The proteins in Freja's skin cells were made using the particular nucleotide sequence of her genes. If the DNA had been altered by UV radiation, different proteins would have resulted.

Hints for Implementing This Task

- Make sure students understand how to make the simulation full-screen.

Possible Extensions

- Have students explore the interactive video "Transcribe and Translate a Gene" at *https:// learn.genetics.utah.edu/content/basics/txtl*.

Student Handout

TASK 21. PROTEIN SYNTHESIS SIMULATION

Introduction

The DNA in the nucleus of a cell contains all the genetic information needed for all the cells in your body. In order for the genetic information to be used, it must first be copied in the nucleus. This copy is then used to make proteins in the ribosomes. DNA is copied into a special molecule known as messenger RNA (mRNA). This process is called transcription. Following transcription, the mRNA is translated at the ribosome into a chain of amino acids that will become the protein. Each group of three nucleotides in the mRNA codes for one amino acid. The protein is built one amino acid at a time until the chain is complete.

Initial Ideas

Before beginning this task, take a few minutes to think about protein synthesis. What is protein synthesis?

How does the sequence of DNA affect the stages of protein synthesis?

Your Task

Your task is to simulate the processes of transcription and translation to see how a gene sequence in DNA can be made into a protein.

Part I: Transcription

1. Access the transcription simulation at *https://authoring.concord.org/activities/22/pages/113/c45b130e-5e22-40d8-bcdc-74210b536a3d*.

2. Scroll down to find the simulation at the bottom right of the website, then click on the *arrows* at *the bottom right corner* to make it full-screen.

3. Click on Prepare for Transcription at the bottom of the graphic. This will separate the DNA strands.

4. Match the complementary RNA nucleotides to the exposed strand of DNA. Start with the first nongreen pair of nucleotides. You'll be done when you reach the red nucleotides.

5. Record the first few RNA nucleotides you have transcribed on the lines above Figure SH21.1 below:

____ ____ ____ ____ ____ ____ ____ ____ ____ ____ ____ ____ ____

Figure SH21.1. Example DNA Sequence

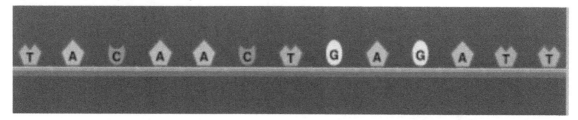

6. Which RNA nucleotide binds with adenine (A) on DNA?

7. Why is an mRNA copy made of DNA?

Part 2: Translation

8. Next, click on the number **3** in the blue square at the top right of the website to access the translation simulation.

9. Scroll down to find the simulation at the bottom right of the website, then click on the *arrows at the bottom right corner* to make it full-screen.

10. Click on Translate Step by Step.

11. Fill out the table below as you work through the simulation.

Translation

Step of translation	Describe what is occurring in this step
  Label the following on the diagram: tRNA, mRNA, ribosome, amino acid	
  Label the following on the diagram: codon, anticodon	

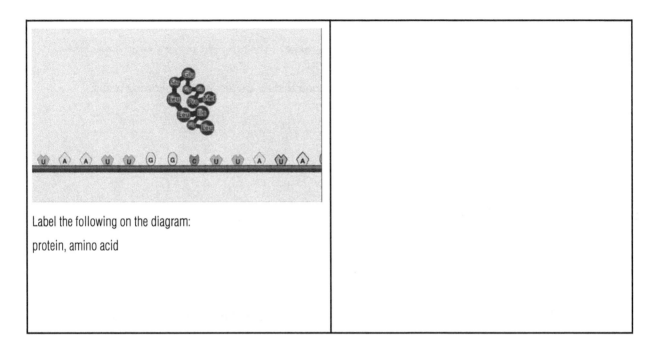

Label the following on the diagram:

protein, amino acid

12. What is the role of tRNA in protein synthesis?

13. How does the mRNA codon determine the amino acid that will be placed into the growing amino acid chain?

Part 3: Protein Folding Simulation

14. Now click on the number **4** in the blue square at the top right of the website to access the protein folding simulation.

15. Read through the information about protein folding and define the following terms:

    a. Hydrophobic

    b. Hydrophilic

16. Scroll down to find the simulation at the bottom right of the website, then click on the *arrows at the bottom right corner* to make it full screen. Follow the directions and fill in the table on the following page.

Protein Folding

	Observations: How does the mutated protein compare with the original protein?	
	Draw the resulting protein (You don't need to label the amino acids, but you should color-code them.)	**Describe** the general shape of the protein
Click on All Hydrophilic and then the Play arrow, and observe what happens to the amino acids.		
Click on All Hydrophobic and then the Play arrow, and observe what happens to the amino acids.		
Click on Mostly Hydrophilic and then the Play arrow, and observe what happens to the amino acids.		
Click on Mostly Hydrophobic and then the Play arrow, and observe what happens to the amino acids.		

17. Explain WHY hydrophilic and hydrophobic amino acids are located in different areas.

Some Useful Ideas From My Teacher

You can keep track of useful ideas from your teacher in the space below.

What We Figured Out

Now that you have completed this task, take a few minutes to fill out the Summary Table below with the other students in your group. This table will help you keep track of what you figured out during the task. You will then have an opportunity to share your ideas as we fill out our class Summary Table.

Summary Table

What we learned	How we know

How it helps us explain the phenomenon

Teacher Notes

Task 22. Mistakes Happen

Purpose

The purpose of this task is to help students visualize the effects that mutations have on proteins. They have already learned that mutations may be caused by environmental factors, such as UV radiation, and that these mutations may occur in tumor suppressor genes and oncogenes, which can lead to a disruption of the cell cycle. They will be able to make connections between what they have learned about cancer, the cell cycle, and protein synthesis and new learning about mutations at the molecular level. This is the culminating task for this unit before students produce a final model and explanation.

Important Life Science Content

DNA mutations can occur because of a substitution, insertion, or deletion of a nucleotide. These mutations may affect the amino acids produced during transcription and the shape of the resulting protein. The mutations can have no effect, a small effect, or a large effect on the amino acids and ultimately the protein.

Scientific Ideas That Are Important to Think About During This Task

- Mutations are a change in the sequence of nucleotides, causing the cell to be genetically different than it was originally.
- The sequence of amino acids in the cell affects the shape of the resulting protein.

Timeline

Approximately one class period.

Materials and Preparation

The items needed for this investigation are listed in Table 3.15. In preparation for this task, we recommend that you familiarize yourself with the online simulation and make sure it can be accessed on student devices.

Table 3.15. Required Materials for Task 22

Item	Quantity
Computer or tablet with internet connection	1 per group
Student Handout	1 per student

Safety Precautions

This task does not require any specific safety precautions.

Procedure

This lesson plan is only a suggestion. It is included here to illustrate how you can facilitate student thinking during this task. We encourage you to modify this lesson plan by asking different questions, using different examples, and providing different scaffolds as appropriate to better meet the needs of students in your class.

Putting Ideas on the Table (50 minutes)

1. Introduce the task with the PowerPoint template we have provided, which can be downloaded from the book's Extras page at *www.nsta.org/mbi-biology*.

2. Students should work in groups through the DNA mutation simulation at *www.biology-corner.com/worksheets/DNA-sim.html*.

3. During the task, be sure to walk around to each group using appropriate Back Pocket Questions (below) to press students to think more deeply about the cell cycle and its relationship to cancer.

4. Once the groups have completed the task, conduct a whole-class discussion to help make sense of the data in relation to Freja's story.

Adding Information to the Summary Table (10 minutes)

1. Give students 5 minutes to decide what to add to the Summary Table at the end of the Student Handout.

2. Have one student from each group share what the group figured out, how they know (their evidence for what they figured out), and how this information will help them explain the anchoring phenomenon.

3. Once each group has shared, ask the entire class to decide what should be added to each column of the class Summary Table for Task 22. Help students reach consensus about what to add to the Summary Table. Only add an idea to the Summary Table if everyone in the class agrees with that idea.

Back Pocket Questions

As students work in groups, it is important to engage with each group to help press and extend students' thinking around the ideas at play in this task. Following are some example questions you might ask:

1. Helping students get started: What patterns do you notice about the different types of mutations and the results of these mutations?

2. Pressing further: How is this related to what you already know about cancer, the cell cycle, and protein synthesis?

3. Following up: What makes you think that? Can you say more?

Filling Out the Summary Table

Table 3.16 includes examples of the responses students may come up with when they fill out the Summary Table. This is provided here only as a behind-the-scenes roadmap and is not meant to be shared with students.

Table 3.16. Example Summary Table for Task 22

What we learned	How it helps us explain the phenomenon
Mutations are caused by a change in the sequence of DNA nucleotides that affects the mRNA nucleotides produced and the amino acids created. The sequence of amino acids affects the shape of the resulting protein.	The mutations that occur in areas responsible for normal functioning of the cell cycle lead to a change in the sequence of the tumor suppressor gene or oncogene, causing mRNA, amino acid sequences, and finally the shapes of proteins to be altered. The proteins can no longer do the same job, resulting in uncontrolled cell division.

Possible Extensions

- Students may explore examples of "The Outcome of Mutation" at *https://learn.genetics.utah.edu/content/basics/outcomes*.

Student Handout

TASK 22. MISTAKES HAPPEN

Introduction

You have learned that cancer is caused by DNA mutations that result in the uncontrolled division of cells. In a normal cell, DNA is transcribed into mRNA and then translated into an amino acid sequence. To understand how Freja's cells were altered, you need to figure out how proteins are synthesized when the DNA has been mutated.

Initial Ideas

Before beginning this task, take a few minutes to think about DNA mutations. What are mutations?

Why does it matter if DNA is mutated?

How does a mutation in DNA affect protein structure?

How does this information link to what you already know about cancer and the cell cycle?

Your Task

Your task is to simulate and explain the effects of DNA mutations on proteins, using an online simulation.

1. Access the DNA Mutation Simulation at *www.biologycorner.com/worksheets/DNA-sim. html* and click on Transcribe.

2. Once the transcription is finished, click on Translate to create the amino acid chain.

3. In the box below, draw the resulting original protein produced. There should be 11 amino acids. Color-code and record the amino acid sequence. (*Hint:* Click on the Stop button to make the model of the protein stop jiggling.)

Original Protein

4. Click on the Edit DNA button. You will now see the original nucleotide sequence used to make the protein. Make the following mutations, one at a time, and then check the new protein created by your new DNA. In the DNA Mutation Analysis table on the next page, describe how each change in the DNA affected the protein produced.

- 1st Mutation → Edit the DNA by changing the *first codon* from (ATG) to (AAG) and click on Apply. Record the changes to the protein in the table, then reset the gene to its original state.

- 2nd Mutation → Change the *second codon* from (CCA) to (CCC) and click on Apply. Record the changes to the protein in the table, then reset the gene to its original state.

- 3rd Mutation → Change the *second to LAST codon* from (TTA) to (TGA) and click on Apply. Record the changes to the protein in the table, then reset the gene to its original state.

- 4th Mutation → Change the *first codon* again by *removing the G*, changing (ATG) to (AT), and click on Apply. Record the changes to the protein in the table, then reset the gene to its original state.

- 5th Mutation → Change the *first codon* yet again by placing an *additional A after the G*, so your strand reads (ATGA), and click on Apply. Record the changes to the protein in the table.

DNA Mutation Analysis

Type of mutation	Location of the mutation (beginning, middle, or end of the gene)	Observations: How does the mutated protein compare with the original protein?	
		Draw the resulting protein	*Describe* similarities to and differences from the original
1st Mutation substitution: missense ATG → AAG			
2nd Mutation substitution: silent CCA → CCC			
3rd Mutation substitution: nonsense TTA → TGA			
4th Mutation frameshift: deletion ATG → AT			
5th Mutation frameshift: insertion ATG → ATGA			

5. How many nucleotides are changed when a substitution mutation occurs? How many amino acids will be affected by a single substitution?

6. Which type of mutations caused the biggest changes to the resulting protein: substitution or frameshift? Why?

7. What does it mean for a mutation to be "silent"?

Some Useful Ideas From My Teacher

You can keep track of useful ideas from your teacher in the space below.

What We Figured Out

Now that you have completed this task, take a few minutes to fill out the Summary Table below with the other students in your group. This table will help you keep track of what you figured out during the task. You will then have an opportunity to share your ideas as we fill out our class Summary Table.

Summary Table

What we learned	How we know

How it helps us explain the phenomenon

Building Consensus Stage Summary

In the building consensus stage of an MBI unit (Figure 3.8), students finalize their group models, compare and contrast those models as a class as they work to reach a consensus, and co-construct a Gotta Have Checklist of the ideas and evidence that should be part of their final evidence-based explanations. Before beginning this stage, review the relevant sections in Chapters 1 and 2 for specifics. See Figures 3.3 and 3.4 (p. 346) for examples of a Gotta Have Checklist and a final model for this unit. We have provided a PowerPoint template for this stage, which can be downloaded from the book's Extras page at *www.nsta.org/mbi-biology*.

Figure 3.8. Building Consensus Stage Summary

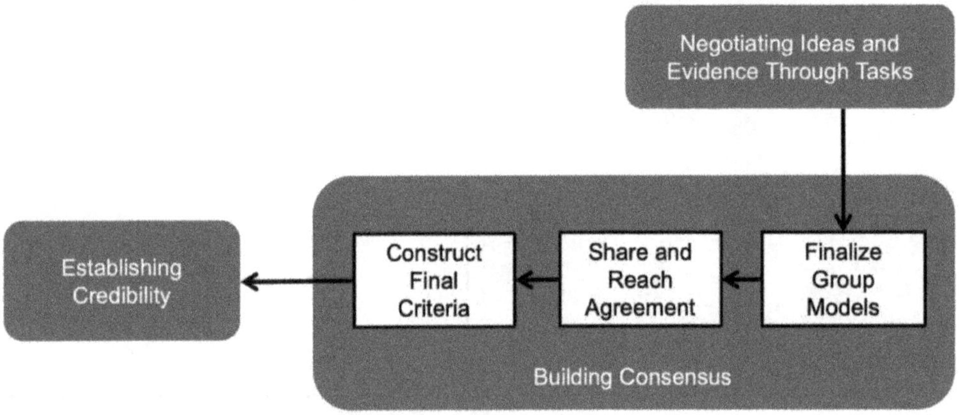

Finalize Group Models

Finalizing the models requires groups to review their previous models, decide what needs to be revised based on new ideas and understandings from the previous set of tasks, and redraw the models so they can be more easily shared and used to build consensus in a whole-class setting. This usually takes about 30 minutes. We often scaffold this process by asking students to review the completed Summary Table and talk about what should be added to, removed from, or changed in their previous models. They should think about what new ideas can help them move toward a truly causal model that includes not only *what* happened but also *why* it happened. This requires that the mechanism at play be visible and well explained. We push the groups to make sure that items in their models are labeled, that any unseen components are made visible, and that the important ideas that surfaced throughout the unit are explicitly used in the models.

Share and Reach Agreement

We think a share-out session is needed here so the groups can learn from each other and begin to build consensus across the models. There are a number of ways to do this, including gallery walks. As the goal is to build consensus as a whole class, we usually opt to facilitate a share-out session in which each group comes to the front of the class, displays its model, and talks it through. We prompt students to ask questions, and the class works hard to compare and contrast ideas across the models.

Construct a List of Final Criteria

The goal of our last public record, the *Gotta Have Checklist*, is to have the class negotiate about which of the main ideas and evidence should be part of a complete and scientifically defensible explanation of the phenomenon. Students will use this as a scaffold for writing their final evidence-based explanations in the last stage of the unit. There are a number of ways to facilitate this discussion and creation of the public record. We like to prompt groups to create a bulleted list of the three to five most important ideas that they think they need. We then ask for examples, press the whole class to make sure they understand how the idea fits in, and ask for consensus before writing it on the final checklist public record. This process can take about 15 to 20 minutes and most often goes quite smoothly by this point in the unit. However, this is also a time to make sure important ideas are included and to bring them up if necessary. Although this is uncommon, we may also provide some just-in-time instruction to tie up any loose ends in students' understandings from the unit.

The checklist becomes less useful to students if everything that comes up in this discussion is automatically written on the board. Ideas can generally be combined into five to seven bulleted points. We then go back and lead a discussion about the evidence we have for each of the bulleted points and write those alongside. This step is crucial, as students will need to coordinate the science ideas with evidence in the written evidence-based explanation in the next stage. At times, we have been unhappy with our checklist for some reason; perhaps the writing is not as legible or the points are not as clearly articulated as we would like. In such cases, we created a clean version that evening and asked the students the next day to ensure that it still represented all the ideas from the original poster.

Establishing Credibility Stage Summary

In the establishing credibility stage (Figure 3.9), students construct arguments in the form of evidence-based explanations of the phenomenon, which allows them to argue their ideas in writing. This stage also includes peer review and revision to establish credibility for their ideas. Before beginning this stage, review the relevant sections in Chapters 1 and 2 for specifics. We have provided a PowerPoint template to assist with this stage, which can be downloaded from the book's Extras page at *www.nsta.org/mbi-biology*.

Figure 3.9. Establishing Credibility Stage Summary

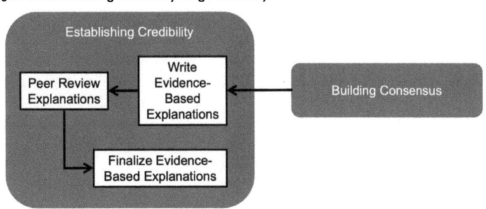

Write Evidence-Based Explanations

Up to this point, students have worked as part of a group to construct their evidence-based explanations of the phenomenon. Now, they work individually to write their final explanations. While they are on their own, they have a number of important public records such as the Summary Table and Gotta Have Checklist to use as references. Beyond these scaffolds, you may choose to provide additional support for students who may struggle with writing. For instance, we often have the first sentence written out on small slips of paper that we can hand to students to help build momentum. Some teachers choose to have student groups work together to create a group outline of their writing before students work individually. Your school may have writing support structures in place to help students get started. One thing we have found is that teachers are most often pleasantly surprised by their students' individual evidence-based explanations.

Conduct Peer Review of Explanations

Once students have finished writing their individual evidence-based explanations, it is important that they receive peer feedback before turning in their final product. Have students use the MBI Explanation Peer-Review Guide in Appendix B to provide feedback to each other. Beyond helping students improve their written explanations, this step reinforces the idea that in science, credibility comes not from an external standard but through negotiation with our peers.

Finalize Evidence-Based Explanations

After students have been able to review their peer feedback, and perhaps meet with their peers to discuss the feedback, they should use the critique to finalize their evidence-based explanations.

Evaluation

Once students have handed in their written explanations, you can use the MBI Explanation Peer-Review Guide in Appendix B to evaluate them. Besides clarity of communication, the guide focuses on whether a student's final product explains the phenomenon, fits with the evidence gathered throughout the unit, and builds on the important science ideas at the center of the unit.

Unit 4. Biological Evolution

Unity and Diversity (LS4): *Lampsilis* Mussel

Unit Summary

Evolution is one of biology's most important and far-reaching theories. The theory of evolution by natural selection ties together the otherwise loose collection of topics in biology, from molecular genetics to paleontology. Here, we focus on the variation of traits, heritability, adaptation, and natural selection as core concepts of evolutionary theory. We define *evolution* as the change in heritable characteristics of biological populations over successive generations. This process is driven by natural selection in response to changes in an organism's environment. We use the *Lampsilis* mussel as the anchoring phenomenon for the unit, as there is an advantage to studying this mussel to learn about natural selection. Students often struggle to understand evolution because they anthropomorphize the organisms under study—for example, cheetahs "want" to evolve faster speed or giraffes "want" to evolve a longer neck. However, the *Lampsilis* mussels lack a brain and even eyes, and they have no ability to see the fish they are mimicking with their lures, which removes anthropomorphism as a possible mechanism and allows students to focus on the important concepts at hand. The unit uses six tasks to support students in learning about these important concepts while using them to build their evidence-based explanations of the evolution of *Lampsilis* mussel lures over time. This unit also draws on what the students learned about heredity while studying skin cancer in Unit 3. The tasks engage students in the disciplinary core idea (DCI) Life Science 4 (LS4): Biological Evolution: Unity and Diversity.

Table 4.1 provides an outline for this unit, describing the purpose and the major tasks and products for each MBI stage, as well as an approximate timeline.

Table 4.1. MBI Outline and Timeline

MBI stage	Purpose	Major tasks and products	Timing
Eliciting ideas about the phenomenon	To introduce the phenomenon, eliciting students' initial ideas to explain the phenomenon, and begin constructing group models.	Initial ideas public record; initial group models	1–2 days
Negotiating ideas and evidence through tasks	To provide opportunities for students to make sense of the phenomenon through purposeful tasks and discussions.	Individual task products; revised group models	7–8 days
Building consensus	For groups to come to an agreement about the essential aspects of the explanation.	Final group models	1 day
Establishing credibility	For each individual student to write an evidence-based explanation.	Final evidence-based explanations	2 days
		Total	11–13 days

Anchoring Phenomenon

Lampsilis mussels, like other mussels, have a complicated life cycle involving a parasitic stage in which microscopic larvae must reside in living hosts to complete the cycle. For the *Lampsilis* mussel, this host is the largemouth bass. This presents a challenge, as the bass do not prey on mussels and therefore have little reason to get close enough for the mussels to deliver their larvae. Amazingly, the mussels have developed lifelike lures to attract the bass. The mussel uses flesh from its mantle that not only looks like the bass's prey but also moves like them as the mussel twitches its lure. When the bass attacks the lure, a membrane is broken, firing the mussel's larvae into the fish's gills. The parasitic larvae attach and drain nutrition from their host before eventually falling off and settling on the riverbed to complete their life cycle. Typically, this species mimics the small fish prey of the largemouth bass. There are four widespread types of lures in the genus, some of which are stronger mimics of the bass prey than others. The most effective mimics are the darter fish lures, some of which even sport eyespots and what look like gasping mouths. These occur at a higher rate in the population. Amazingly, the mussels have developed these lifelike lures even though they have no eyes and have never seen the fish they are mimicking.

Driving Question

How did *Lampsilis* mussels develop lifelike lures over time?

Target Explanation

Following is an example evidence-based explanation that could be expected at the end of the unit. We consider this an exemplar final explanation at this grade level. This is included here to help support you, the teacher, in responsively supporting students in negotiating similar explanations by the end of the unit. It is not something to be shared with students but is only provided as a behind-the-scenes roadmap for you to consider before and throughout the unit to guide your instruction.

> *Lampsilis* mussels have a complex life cycle. Males release sperm into the water to fertilize the females' eggs, which develop into larvae within modified gill structures. To complete the cycle, the female must deliver her larvae into the gills of a largemouth bass. In the bass, the larvae snap onto the gills and act as parasites, gaining protection and feeding off of the fish's blood until they are sufficiently developed, at which point they drop off and continue their development to adulthood on the riverbed, ready to complete their life cycle.
>
> Delivering the larvae into the bass's gills is a challenge. Mussels cannot swim and bass are uninterested in mussels. Simply releasing their young into the water is ineffective, as few reach the bass's gills. Instead, the *Lampsilis* mussels have evolved lifelike mantle lures to attract their hosts. These lures are adaptations, or traits that make the mussels more successful in their environment, as the lures increase the likelihood of the larvae completing the life cycle. The bass has a voracious appetite for small fish, especially darter fish, and the lures mimic these small fish prey. When a bass attacks a mussel's mantle lure, the mussel's gills rupture and release the larvae to be taken up by the bass's gills. These adaptations first

occurred because of mutations, or changes to the DNA, in the mussels. We saw how changes in DNA affect the amino acid sequence and finally the shape of the protein in the Lure Morphology task.

Once the adaptation existed, genetic recombination has worked on it to provide a number of different variations. We saw this in the field mice data from the online simulation in the Genetic Recombination task. The mussels have evolved four types of lures that vary in how well they mimic, or look like, the prey of largemouth bass, according to data collected from the Great Lakes drainages of Ontario. The most effective lure looks remarkably like the darter fish, a small prey fish of the bass. Within the population of mussels, there is a lot of variation in the lures. Some even have fake eyespots and mouths and are able to perfectly mimic darter fish movements. Incredibly, although mussels have no eyes and have never seen a darter fish, the lures have evolved over millions of years.

Lures that best resemble prey fish of largemouth bass are most advantageous to the mussels, as bass are more likely to mistakenly attack them as prey. We saw evidence of this in the I Will Survive! task. The traits that positively affected survival rates were more likely to be reproduced and thus become more common in the population. This is an example of natural selection. Those individuals most fit for their environment (i.e., those with lures that look most like the darter fish) survive and go on to reproduce offspring that have those advantageous traits as well. The genes that create the most successful lure are passed down, while the genes for those that were unsuccessful are not. Over time, this can change the entire population of mussels, as those with the most successful fishlike lures are more often attacked by bass and reproduce more than others. Over many generations, the bass's preference for darter fish has acted as a selection pressure on the mussels, shaping the current and future populations. This is similar to what we learned in the Galapagos Finches and Does Beak Size Matter? tasks. Because of changes to the environment, the ratio of small to large seeds changed on the island of Daphne Major, which led to changes in the beak sizes of the local finches.

Today, the darter lure is by far the most common, with the three other types of lures less common in the mussel population. In the future, any changes to the environment, such as to the prey species of the bass, will further shape the lures. Changes may even end up producing more species of mussels, a process known as speciation. As we saw in the Where'd Those Mussels Come From? task, the more types of native fish species, the greater the number of mussel species.

Eliciting Ideas About the Phenomenon

The eliciting ideas about the phenomenon stage of MBI is about introducing the phenomenon and driving question, eliciting students' initial ideas about what might explain the phenomenon and answer the driving question, and constructing their initial models in small groups. We have provided a PowerPoint for this purpose that will lead you step-by-step through the process, which can be downloaded from the book's Extras page at *www.nsta.org/mbi-biology*. There are two products of this stage: an Initial Hypotheses List and student groups' initial models. Examples of each are shown in Figures 4.1 and 4.2.

Figure 4.1. Example Initial Hypotheses List

Initial Hypotheses List

- Using senses (other than sight), analyzed predator/prey relationships to create its own 'prey' to manipulate the Large Mouth Bass (LMB)
- Over time, Lampsilis took in DNA from prey fish to adopt the phenotype of those fish
- A long ago, a mussel may have had something resembling a lure which then reproduced creating more mussels with lures
- At point Lampsilis had eyes (& could see prey fish) therefore was able to know what they looked like (no longer has eyes)
- In the past Lampsilis had different hunting strategies, but due to change in environment it caused them to lose some abilities to be replaced with new/different ones
- Lampsilis mussels with lures closest in coloration to prey fish were able to spread their young

Figure 4.2. Example Initial Model

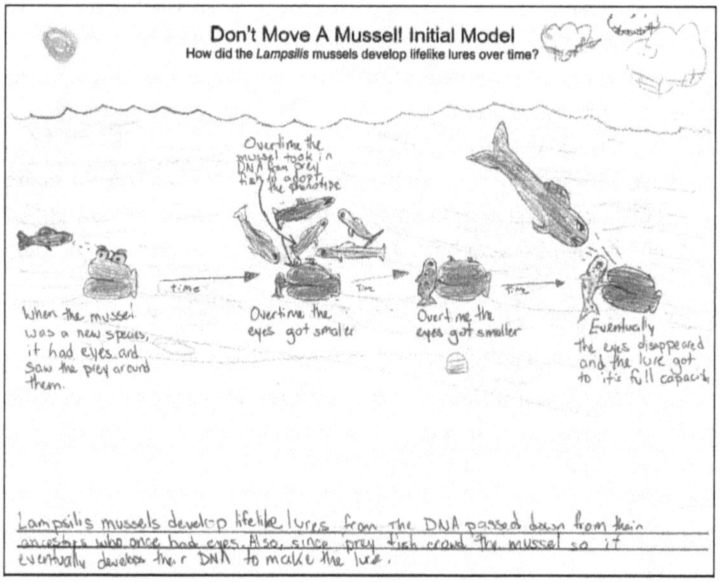

This stage of MBI is critically important, as it orients students to the MBI unit anchoring phenomenon using the driving question in a way that makes it compelling enough to spend several weeks across a unit working on explaining an event. Put more succinctly, this stage introduces the problem space in which the students will be working. Introducing the phenomenon should be done in a way that supports students in drawing on what they have learned previously in and outside of school as they begin to think about how to explain the driving question. Your priority should be to elicit students' ideas about the phenomenon, since it is important for them to continually think about the phenomenon and refine their everyday ways of thinking. Since eliciting students' ideas is a top priority, it's important to create a learning environment where students feel comfortable and are invited to offer ideas. This means that you need to think carefully about how to get students to float (or put on the table for consideration) as many ideas as possible. At this stage, there are no right or wrong ideas. Everything (within reason!) is on the table.

Negotiating Ideas and Evidence Through Tasks

The negotiating ideas and evidence through tasks stage of MBI makes up the majority of the unit. Each task is designed to introduce a key scientific idea or to reinforce an idea already raised by students in the class. Halfway through the unit, we recommend having groups revise their models and share out those revisions. PowerPoint templates are provided for the tasks as well as for the model revision; these can be downloaded from the book's Extras page at *www.nsta.org/mbi-biology*. Each task begins by introducing ideas for students to reason with while working on the task. During the task, students investigate or test the concepts using data. We recommend using the Back Pocket Questions provided in the Teacher Notes to press students' thinking while they are engaged in the task. Taken together, the tasks scaffold students' thinking as they co-construct a scientific explanation of the phenomenon. The general outline of this stage is as follows:

Task 23. Lure Morphology

Task 24. Genetic Recombination

Task 25. I Will Survive!

Midunit Model Revision for Unit 4

Task 26. Galapagos Finches

Task 27. Does Beak Size Matter?

Task 28. Where'd Those Mussels Come From?

A Summary Table is used after each task to scaffold students' negotiation of ideas and how those ideas help explain the phenomenon. An example is provided in Table 4.2 (p. 324). Like the target explanation, this is included here to serve as a behind-the-scenes roadmap to help support you, the teacher, in responsively supporting students in negotiating similar responses by the end of the unit. It is not something to be shared with students but is provided for you to consider before each task to guide your instruction.

Table 4.2. Example Summary Table for Unit 4

Task	What we learned from this task	How it helps us explain the anchoring phenomenon
23. Lure Morphology	Mutations produce new variations in populations. Variations that help an organism survive in its environment are adaptations.	Mutations produce lifelike lure variations. These variations are helpful adaptations if they make the lure appear more lifelike to the host fish.
24. Genetic Recombination	Recombination is a process in which genes are mixed into new combinations that result in variation of traits. Some combinations of genes create more fit, better-adapted individuals than others, which may lead to certain traits becoming more common in a population.	Recombination has led to a number of different variations in mussel lures. Some are more successful than others and thus are more common in the population.
25. I Will Survive!	Variation leads to different survival rates among individuals. Traits that positively affect survival are more likely to be reproduced and thus become more common in the population.	There is variation in *Lampsilis* mussel lure morphology. Mussels with lures that best mimic the prey species of a host fish are more likely to survive and reproduce, and thus these lures become more common in the population.
26. Galapagos Finches	Evolution is the change in characteristics over time. Changes in the environment can cause large changes in populations, even resulting in new species.	One species of *Lampsilis* mussel diversified into many different species by natural selection.
27. Does Beak Size Matter?	Natural selection is the process whereby organisms better adapted to their environment tend to survive and produce more offspring. This can lead to the creation of new species, known as speciation.	As the environment around the mussels changed, natural selection acted on the variations of mussel lures. As some of these populations were separated, new species arose.
28. Where'd Those Mussels Come From?	Speciation most often occurs because of a physical barrier that impedes the genetic flow of traits between populations.	Host fish are attracted to different prey. *Lampsilis* mussels mimic this prey with their lifelike lures, thereby giving rise to an increase in lure diversity.

You can think of your role here as that of helping introduce ideas to students to support them in learning about and beginning to engage with the ideas through carefully designed sensemaking tasks. If important ideas are not introduced by the students, you can put them on the table or introduce them through the use of just-in-time direct instruction, short videos, or readings, among other strategies. Since the objective is to have students pick up and try out these ideas in carefully crafted tasks as they develop explanations, introducing your students to an idea early in the task is *not* giving them answers that will be confirmed in the task. Instead, you can think of this as introducing students to tools (i.e., disciplinary core ideas) that will be useful in a task in ways that will help them make sense of data, a simulation, an activity, or something in the world. In the end, it is important that you think of introducing new ideas in the early parts of this MBI stage as an opportunity for students to engage in sensemaking with new ideas to use, while you concurrently recognize the potential explanatory power of the new ideas. At the end of each task, the completion of a row of the Summary Table is an opportunity for students to think about what they have learned in the task and, together with peers, reason about how the newly introduced ideas can be applied to support their objective of explaining the unit anchoring phenomenon.

Building Consensus

In the building consensus stage of an MBI unit, the whole class works to build consensus about the explanation of the phenomenon by finalizing the groups' models, comparing and contrasting those models as a whole class, and constructing a consensus checklist of the ideas and evidence (called the Gotta Have Checklist) that should be a part of students' final evidence-based explanations that make up the summative assessment of the unit. A PowerPoint is provided as a guide through this stage of the unit and can be downloaded from the book's Extras page at *www.nsta. org/mbi-biology*. Examples of the whole-class Gotta Have Checklist and a group's final model appear in Figures 4.3 and 4.4 (p. 326).

Figure 4.3. Example Gotta Have Checklist

Gotta Have Checklist

- Male *Lampsilis* mussels release **sperm** into the water, females receive the sperm to fertilize the eggs and **brood** until they release the **glochidia** into the **gills** of the **largemouth bass** where they **attach** themselves until they are mature enough to **release** and fall to the riverbed and **anchor** themselves in the sand
- **Female** *Lampsilis* have a lifelike lure to **attract** largemouth bass close enough that likely originated as an accidental **chromosomal duplication mutation** creating an **extension of the mantle** tissue
- The extension of the mantle was an **advantage** allowing Lampsilis to **reproduce more often** through natural selection passing its **advantageous genes** on to the **next generation**
- The **length** of the mantle extension (lure) **increased over time** due to **directional selection** as longer extensions were **more advantageous** and added more lifelike features (such as **eyespots** & **darter fish** patterning) through **additional mutations** that also **increased likelihood** of **survival** and **reproduction**
- Mussels utilize 3 sensory receptors (**chemoreceptors**, **mechanoreceptors**, and **photoreceptors**) to help them **detect the presence** of the largemouth bass in order to create **lifelike movement** of their lure

Figure 4.4. Example Final Model

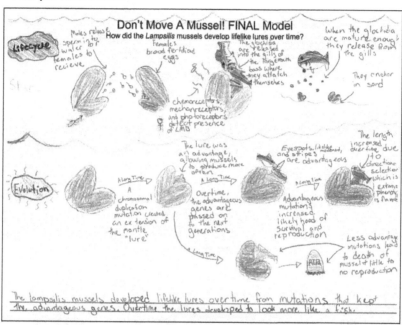

In Chapter 2, we described the scientific practice of modeling as a knowledge-building practice and models as the products of modeling. Students begin this stage by finalizing their models in groups of three or four, which they should think of as a continuation of the sensemaking that started in the groups' initial modeling experiences in the early days of the MBI unit and continued when they revisited these models midway through the unit. As with these other experiences, you can expect the practice of modeling to support groups in refining their explanations of the unit anchoring phenomenon. Here, negotiation and argumentation lead to refined ways of thinking about the phenomenon that may not have occurred had students not engaged in modeling this final time. The groups' models will offer insight into your students' thinking that otherwise may not be accessible to you. At the same time, models as products of modeling also serve as artifacts that students can compare across groups in ways that are supported by your selection and sequencing of groups' sharing in whole-class sensemaking, so that differences can be foregrounded, negotiated, and resolved through argumentation, among other science practices.

Finally, the construction of final criteria or whole-class mapping of ideas included in the Gotta Have Checklist to the evidence collected from the various tasks completed in the second stage of MBI is a scaffold that will be important for individual students in writing evidence-based explanations. Importantly, the teacher-facilitated, co-constructed Gotta Have Checklist outlines what the class agrees should be included in students' explanations and is mapped to evidence identified across the unit tasks. This step is the last stage of building consensus. This is especially true because final group models will have been negotiated in both small and whole groups as a basis for the identification of the criteria that are needed for writing an evidence-based explanation.

Establishing Credibility

In MBI, students must argue for their ideas in writing. In the establishing credibility stage, they do this through written evidence-based explanations, peer review, and revision. A PowerPoint is provided as a guide through this stage of the unit and can be downloaded from the book's Extras page at *www.nsta.org/mbi-biology*. Writing scaffolds such as sentence stems may be necessary to jumpstart some students' writing. Once the written explanations are complete, we recommend using the MBI Explanation Peer-Review Guide in Appendix B. The target explanation on pages 320–321 is an example of a written explanation.

Writing evidence-based explanations is a sensemaking experience that students engage in early in this stage of MBI. This is another knowledge-building opportunity that, like engaging in modeling earlier in the unit, is a practice that leads to a product. And like models, which served as products that could provide you with insight into the way groups were thinking about explaining the anchoring phenomenon, the individually written evidence-based explanations afford you with an opportunity to assess individual students in terms of where they are in their attempts to explain the anchoring phenomenon. As students engage with others in this stage, this is an additional space for argumentation and negotiation as they recognize possible differences and details they may not have considered or may not yet agree with.

This practice of peer review needs to be supported, so we have provided the MBI Explanation Peer-Review Guide in Appendix B, which students can use to provide feedback to each other. The peer-review guide contains questions designed to focus student attention on the most important aspects of explanations and to encourage students not only to discuss what they figured out and how they know what they know but also to best communicate what they have learned to others in a way that is clear, complete, and persuasive. During the peer review, you should consider and make explicit how explanations, like models, are refined over time through negotiation within a community. The MBI unit culminates at the end of this stage as students use the feedback they have received to make final revisions that represent how they are thinking about the unit anchoring phenomenon. The final evidence-based explanations serve as one measure of what students have learned about engaging in science practices to use disciplinary core ideas and crosscutting concepts to explain the phenomenon.

Hints for Implementing the Unit

- You may wish to begin this unit with a mussel dissection and anatomy lesson. An example lesson can be found at *www.shapeoflife.org/sites/default/files/SoL-mussel-lesson-Middle-School_0.pdf*.

- For additional information and resources concerning evolution, see "Understanding Evolution" at *http://evolution.berkeley.edu*.

- We recommend providing opportunities to connect students' personal experiences and community resources with the phenomenon and topic of this unit. While students may not have had any personal experiences with the phenomenon, we suggest holding discussions throughout the unit to elicit their personal connections. Think about possible connections in your community that you can bring to the conversation as well. These kinds of connections will make the unit personally meaningful to your students, increasing their motivation to engage in the tasks. In our experience, much of this happens during the eliciting ideas about the phenomenon stage of the unit. However, you should prompt students to connect to the topic during all phases of the unit.

- There may be times during the unit when students make connections to injustices or disparities in their communities or raise emotional responses to the topics. We suggest preparing to address young people's questions and desire for activism so that science practices are not portrayed as disconnected from the social and cultural contexts of students' real-world experiences.

Targeted NGSS Performance Expectations

- **HS-LS4-1.** Communicate scientific information that common ancestry and biological evolution are supported by multiple lines of evidence.

- **HS-LS4-2.** Construct an explanation based on evidence that the process of evolution primarily results from four factors: (1) the potential for a species to increase in number, (2) the heritable genetic variation of individuals in a species due to mutation and sexual reproduction, (3) competition for limited resources, and (4) the proliferation of those organisms that are better able to survive and reproduce in the environment.

- **HS-LS4-3.** Apply concepts of statistics and probability to support explanations that organisms with an advantageous heritable trait tend to increase in proportion to organisms lacking this trait.

- **HS-LS4-4.** Construct an explanation based on evidence for how natural selection leads to adaptation of populations.

- **HS-LS4-5.** Evaluate the evidence supporting claims that changes in environmental conditions may result in: (1) increases in the number of individuals of some species, (2) the emergence of new species over time, and (3) the extinction of other species.

- **HS-LS4-6.** Create or revise a simulation to test a solution to mitigate adverse impacts of human activity on biodiversity.

Eliciting Ideas About the Phenomenon Stage Summary

The first stage of MBI, eliciting ideas about the phenomenon (Figure 4.5), involves introducing the anchoring phenomenon and driving question, eliciting students' initial ideas and experiences that may help them develop initial explanations of the phenomenon, and developing initial models of the phenomenon based on those current ideas. Before beginning this stage, review the relevant sections in Chapters 1 and 2 for specifics. See Figure 4.1 on page 322 for an example of an Initial Hypotheses List for this unit. We have provided a PowerPoint to assist with this stage, which can be downloaded from the book's Extras page at *www.nsta.org/mbi-biology*.

Figure 4.5. Eliciting Ideas About the Phenomenon Stage Summary

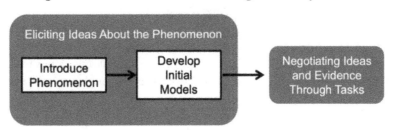

Introduce the Phenomenon

We begin this stage by introducing the phenomenon in an engaging way, such as with stories, videos, demonstrations, or even short activities. The goal is to provide just enough information for students to begin to reason about the phenomenon, without providing too much of the explanation. While we are introducing the phenomenon, we ask questions to keep students engaged and make sure they are paying attention to the important aspects that can begin to help them explain the phenomenon. The introduction ends with the driving question of the unit, which we have found helps focus their thinking on the development of a causal explanation for the phenomenon, or a "why" answer.

It is important to spend time eliciting not only students' scientific ideas about the phenomenon but also their personal connections to the phenomenon. What personal experiences do they have that help them connect to the phenomenon? While the phenomenon of the unit may not be directly connected to your community, does the community have similar phenomena or specific resources you can bring into the discussion? The more connected students feel to the unit phenomenon, the more engaged and motivated they will be.

We then get students into their groups and facilitate the first discussion to try to answer the driving question with just the resources they brought with them—the ideas, experiences, and cultural resources they have gained both in and outside of the classroom. These ideas may be fully formed, partially correct, or fully incorrect in terms of our canonical knowledge of science. However, we make it clear that all ideas are considered equally valid at this point in time, as we realize that these are the ideas students put into play when they think about the phenomenon we have introduced.

Once student groups have discussed their ideas, we facilitate a class discussion to compare and contrast the ideas generated by each group. As ideas are presented, they are put on our first public record, which we call the Initial Hypotheses List. We use this list throughout the unit to keep track of the changes in students' thinking as they work toward a final evidence-based explanation for the anchoring phenomenon. We consider all ideas to be valid at this stage, before students use other resources to begin making sense of how or why something happens. The Initial Hypotheses List is also useful for the next task in this stage, initial model construction, especially since it offers students additional ideas beyond those they initially had either individually or in their small groups.

Develop Initial Models

If students are not experienced with modeling, it is worth providing a brief introduction and example. We have included an example in the PowerPoint for this stage, which can be downloaded from the book's Extras page at *www.nsta.org/mbi-biology*. Once the class is ready to begin modeling, we give each group a sheet of 11 × 17 inch paper and ask the groups to each make a model of their initial hypothesis. Sometimes they choose their own original hypotheses, and other times they are influenced by their peers' ideas and adopt one of them instead. As the groups work on constructing their models, you should walk around asking clarifying questions and pushing students to be as specific as possible. Once the models are ready, it is important to have students share ideas across them. There are a number of ways you might run these share-out sessions. We often collect and present the models on a document camera at the end of the first day. Groups can provide one- or two-sentence summaries of the initial hypotheses they have represented in their models. We point out interesting ideas and ways in which they have represented these ideas. For example, we may call attention to the fact that a group labeled the arrows, which made the model more understandable, and that another group used a zoom-in window to show what was happening at a different scale. At the end of the first day and this first stage, we have elicited ideas across the class, and the groups' initial hypotheses and models will act as a starting point for the rest of the unit.

Negotiating Ideas and Evidence Through Tasks Stage Summary

The next stage of MBI, negotiating ideas and evidence through tasks (Figure 4.6), takes up the majority of the unit. The tasks are designed to introduce or extend important science ideas that students need as they construct their evidence-based explanations of the phenomenon. Before beginning the tasks, we suggest you review the relevant sections in Chapters 1 and 2 for specifics. Each task consists of the following:

- A **Teacher Notes** section that provides an overview of the activity and science content important to the task. Here, we provide guidance on conducting the task with details on the procedure, a list of required materials and preparation, any necessary safety precautions, suggested Back Pocket Questions, and an example Summary Table entry. At the end of this section are further hints for implementing the task and possible extensions if desired.

- A **Student Handout** to be given to groups as they engage in the task. The handout provides an introduction to the task and content, an initial ideas section to frame the learning before they begin, and detailed instructions and work space. It ends with a section called What We Figured Out, which is designed to scaffold student responses to include in the Summary Table as a whole class. We suggest having students fill out this section after completing the task and the post-task discussion.

- A **PowerPoint** Negotiating Ideas and Evidence Through Tasks template that you can adapt for each task, which can be downloaded from the book's Extras page at *www.nsta.org/mbi-biology*.

- A Midunit Model Revision reminder roughly halfway through the unit.

Figure 4.6. Negotiating Ideas and Evidence Through Tasks Stage Summary

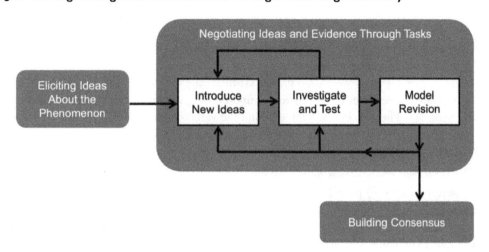

Teacher Notes

Task 23. Lure Morphology

Purpose

The purpose of this two-part task is for students to review the concept of mutation as examined in the skin cancer unit and then connect this to the concepts of variation and adaptation. In Part 1, students review the steps involved in protein synthesis and determine that mutations that occur during this process result in the formation of unexpected proteins or traits, which can produce genetic variation. Then, in Part 2, they examine the variations in the mussels' lure adaptations through data on lure morphology from the Great Lakes drainages.

Important Life Science Content

DNA sequences (genes) provide instructions to make proteins, and proteins are the building blocks of organisms. Mutations are permanent changes that occur in the DNA sequence and include insertions, deletions, frame shifts, and duplications. They may affect only one nucleotide (A, C, G, or T) or whole sections of chromosomes. They occur as a result of either mistakes when the DNA is copied or environmental factors such as UV radiation or cigarette smoke. These changes are often detrimental to the organism but are sometimes neutral or even advantageous. As DNA provides instructions for proteins, mutations in DNA affect the resulting proteins. Those mutations that are advantageous result in variation within species. For example, the disorder sickle cell anemia is caused by a mutation in the gene that instructs the building of a protein called hemoglobin. This causes the red blood cells to become an abnormal, rigid sickle shape. However, in African populations, having this mutation also protects against malaria. Cancer is the most common human genetic disease; it is caused by mutations occurring in a number of growth-controlling genes (see Unit 3). Mutations that help a species survive in its environment lead to adaptations. In the anchoring phenomenon for this unit, the lifelike lure of the *Lampsilis* mussel is an adaptation, caused by a mutation in the evolutionary history of the species, that increases its chances of surviving and reproducing in its current environment.

Scientific Ideas That Are Important to Think About During This Task

- A mutation is a change that occurs in the DNA sequence as a result of either mistakes when the DNA is copied or environmental factors.
- Mutations produce new variations in traits.
- Variations are the differences in genes, traits, or behaviors among members of a population and may result in differences in reproductive success.
- Variations that help an organism survive in its environment lead to adaptations.
- Adaptations are traits that help an organism survive in its environment.
- The environment determines which variation of a trait will help an organism survive.

Timeline

Approximately two class periods.

Materials and Preparation

The items needed for this investigation are listed in Table 4.3.

Table 4.3. Required Materials for Task 23

Item	Quantity
Colored pencils	1 set per group
Crayons	1 set per group
Marker pens	1 set per group
Student Handout	1 per student

Safety Precautions

This task does not require any specific safety precautions.

Procedure

This lesson plan is only a suggestion. It is included here to illustrate how you can facilitate student thinking during this task. We encourage you to modify this lesson plan by asking different questions, using different examples, and providing different scaffolds as appropriate to better meet the needs of students in your class.

Introduction to the Task (10 minutes)

1. Before introducing this task, be sure to review the anchoring phenomenon and students' list of initial hypotheses with the class. Remind students that all tasks completed during this unit are intended to provide them with scientific ideas and evidence they will use to explain the phenomenon.

2. Introduce the task by reinforcing these two main ideas: (1) mutations produce new variations in traits, and (2) mutations that produce traits that help an organism survive in its environment are called adaptations.

3. Lead students through a transcription/translation practice problem.

Part 1 (50 minutes)

1. Pass out the Part 1 Student Handout. Students should work in groups to transcribe and translate each gene, then use the resulting amino acid sequence to draw a lifelike lure. Next, students consider how a mistake during the process of protein synthesis might alter the phenotype of the lifelike lure and whether the resulting lifelike lure variation would increase or decrease the mussels' chances of reproducing.

2. Review students' work as a class. Have students share their lifelike lure drawings, including traits resulting from harmful and beneficial mutations they created. Emphasize that mutations happen all the time and produce variation, and that it is the environment (host fish) that determines whether the variation is beneficial (an adaptation) or not.

Part 2 (50 minutes)

1. Pass out the Part 2 Student Handout.

2. Students should work in groups to apply what they have learned about natural selection to a case study examining lure morphology from *Lampsilis* mussels collected from the Great Lakes drainages of Ontario, Canada (Zanatta, Fraley, and Murphy 2007). Researchers collected more than 75 *Lampsilis* mussels and identified three lifelike lure morphologies. One morphology, the darter-like lure, was significantly more common than either the orange lure or the hellgrammite-like lure. Describe this case study to students, and then have them work in groups to explain the variation seen in this adaptation, which lure is the most common, and why.

Adding Information to the Summary Table (10 minutes)

1. Give students 5 minutes to decide what to add to the Summary Table at the end of the Student Handout.

2. Have one student from each group share what the group figured out, how they know (their evidence for what they figured out), and how this information will help them explain the anchoring phenomenon.

3. Once each group has shared, ask the entire class to decide what should be added to each column of the class Summary Table for Task 23. Help students reach consensus about what to add to the Summary Table. Only add an idea to the Summary Table if everyone in the class agrees with that idea.

Back Pocket Questions

As students work in groups, it is important to engage with each group to help press and extend students' thinking around the ideas at play in this task. Following are some example questions you might ask:

1. Helping students get started: How does the sequence of DNA relate to proteins?

2. Pressing further: How would you describe what happens during a mutation in terms of the ACGT sequence? How do these mutations in DNA sequences relate to the mussels' lifelike lures? What patterns are you seeing in terms of the different lure morphologies?

3. Following up: What makes you think that? Can you say more?

Filling Out the Summary Table

Table 4.4 includes examples of the responses students may come up with when they fill out the Summary Table. This is provided here only as a behind-the-scenes roadmap and is not meant to be shared with students.

Table 4.4. Example Summary Table for Task 23

What we learned	How it helps us explain the phenomenon
Mutations produce new variations in populations. Variations that help an organism survive in its environment are adaptations.	Mutations produce lifelike lure variations. These variations are helpful adaptations if they make the lure appear more lifelike to the host fish.

Hints for Implementing This Task

- Students always enjoy playing "telephone," and this could be a useful game in reviewing protein synthesis and the mistakes (mutations) that inevitably occur.
- Be sure to reference the Part 1 task when presenting the term *variation*. Students need to make the connection among mutations, variation, and adaptations.

Possible Extensions

- There are many useful animations and simulations of genetic mutations that can be used as needed, such as *https://learn.genetics.utah.edu/content/basics/outcomes*.

Reference

Zanatta, D. T., S. J. Fraley, and R. W. Murphy. 2007. Population structure and mantle display polymorphisms in the wavy-rayed lampmussel, *Lampsilis fasciola* (Bivalvia: Unionidae). *Canadian Journal of Zoology* 85 (11): 1169–1181.

Student Handout

TASK 23. LURE MORPHOLOGY (PART 1)

Introduction

In the unit on skin cancer, you learned about the significance of DNA as the instructions or recipe for assembling proteins into living organisms. You transcribed and translated DNA sequences into proteins and modeled what can happen when unexpected changes, known as mutations, occur. Building on this knowledge, in Part 1, you will learn about genetic variation by transcribing and translating DNA sequences (genes) and using genetic codes to assemble the amino acids into a protein, then determining the phenotype each protein represents and the mussel lure that results. You will also consider how a mistake during protein synthesis might alter the lure and whether the resulting variation would be beneficial or not. In Part 2, you will examine a case study to determine which lure is most common and why, based on the concepts of variation and adaptation.

Initial Ideas

Before beginning this task, take a few minutes to think about the transcription and translation of DNA. How is DNA structured?

Describe or draw the processes of transcription and translation.

Your Task

In Part 1, your task is to transcribe and translate DNA to create proteins, then determine the mussel lure each protein represents. You will also consider how other mistakes during protein synthesis might alter the lure and what that would mean.

Transcribe and Translate DNA Sequences

Transcribe and translate the following five DNA sequences (genes) and use the genetic codes in Figure SH23.1 (p. 338) to assemble the amino acids into a protein. Then use the key provided in Table SH23.1 (p. 338) to determine the phenotype each protein represents and draw the resulting lure onto the mussel provided.

Gene 1

DNA	A T C G C G T T A C C G
mRNA	
tRNA	
Amino acids	
Phenotype	

Gene 2

DNA	C A G G T G A T A C C C
mRNA	
tRNA	
Amino acids	
Phenotype	

Gene 3

DNA	T T A G G C C T A C G G
mRNA	
tRNA	
Amino acids	
Phenotype	

Gene 4

DNA	G G A G G C T T C C G G
mRNA	
tRNA	
Amino acids	
Phenotype	

Gene 5

DNA	T A A C G A A T G C C T
mRNA	
tRNA	
Amino acids	
Phenotype	

Figure SH23.1. Genetic Codes

Table SH23.1. Key

Genes	Amino acid sequence	Trait
Gene 1: body style	Ile - Ala - Leu - Pro	Oval and plump
	Ile - Ala - Phe - Pro	Oval and skinny
Gene 2: tail	Gln - Val - Ile - Pro	Triangular tail
	Gln - Val - Pro - Ile	Square tail
Gene 3: body pigment	Leu - Gly - Leu - Arg	Blue pigment (skin)
	Leu - Gly - Leu - Val	Red pigment (skin)
Gene 4: eyes	Gly - Gly - Phe - Phe	Three eyes
	Gly - Gly - Phe - Arg	Two eyes
Gene 5: mouth	Stop - Arg - Met - Pro	Circular mouth
	Stop - Arg - Met - Glu	Rectangular mouth

Draw the Lure

Draw the lure that matches the phenotypes determined previously in the rectangular box attached to the mussel below.

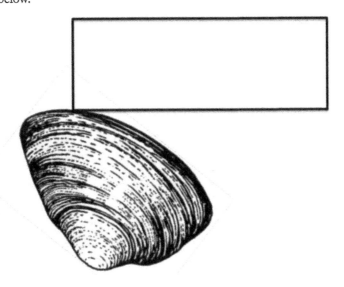

Mutation Questions

What would happen to the lure if a mistake occurred during DNA replication, transcription, or translation? Would the resulting mutation be beneficial or harmful? How would you know? Include an example of a beneficial mutation and an example of a harmful mutation in a new lure drawing below. Be sure to label and describe each below.

What determines whether a mutation is beneficial or harmful for a *Lampsilis* mussel?

Some Useful Ideas From My Teacher

You can keep track of useful ideas from your teacher in the space below.

Student Handout

TASK 23. LURE MORPHOLOGY (PART 2)

Your Task

In Part 2, your task is to read the case study below and examine the related figures, then fill out the totals in the bottom row of the data table and answer the question below the table.

Case Study

In this case study, researchers examined lure morphology in *Lampsilis* mussels collected from the Great Lakes drainages of Ontario, Canada. More than 75 *Lampsilis* mussel specimens were collected in the area outlined by the black box in Figure SH23.2.

Figure SH23.2. Map of Sample Sites

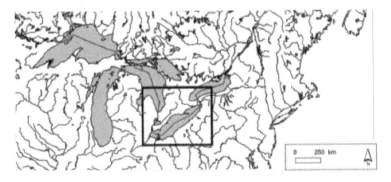

Figure SH23.3 shows three lifelike lure morphologies (variations) found in female *Lampsilis* mussels collected in the study area: (A) orange, no appendages, no eyes; (B) hellgrammite-like, dark, no eyes; (C) darter-like, spots, simple appendages, distinct eyespot. The numbers collected of each of these three variations are given in Table SH23.2 (p. 342).

Figure SH23.3. Diversity of *Lampsilis* Mussel Lure Displays in Great Lakes Drainages of Ontario, Canada

Table SH23.2. Lure Morphologies in Three Great Lakes Drainages of Ontario, Canada

River drainage	Lure morphology		
	Orange lure (image A)	Hellgrammite-like (leopard print) lure (image B)	Darter-like lure (image C)
Grand	3	3	26
Maitland	2	0	9
Thames	6	0	26
Total			

Which lure is most common and why? Use the concepts of variation and adaptation in your answer.

Some Useful Ideas From My Teacher

You can keep track of useful ideas from your teacher in the space below.

What We Figured Out

Now that you have completed this task, take a few minutes to fill out the Summary Table below with the other students in your group. This table will help you keep track of what you figured out during the task. You will then have an opportunity to share your ideas as we fill out our class Summary Table.

Summary Table

What we learned	How we know

How it helps us explain the phenomenon

Teacher Notes

Task 24. Genetic Recombination

Purpose

In the previous task, students learned about mutation as an important source of variation. However, in eukaryotic organisms, those that have cells with a nucleus and other organelles, not only does mutation create new variations of a gene or genes, but during sexual reproduction, these genes are randomized into sex gametes and then recombined into new, random assortments every generation, providing another important source of variation. In this task, students use an online simulation of deer mice to examine the impact of genetic recombination on the variation of traits within a population. Students breed pairs of mice with different fur colors and investigate the ratios of fur color phenotypes, genotypes, and sexes among the offspring of a variety of parent mice combinations. The simulation helps demystify the science behind Punnett squares and encourages students to explore data and statistical representations in genetics and heredity.

Important Life Science Content

Although mutation is the only way of producing new variants of single genes, recombination, brought about through sexual reproduction, produces more new types of individuals much faster than mutation. In eukaryotic organisms, therefore, recombination is the greatest source of variation. In this process, pieces of DNA are broken and recombined to produce new combinations of alleles. This recombination process creates diversity at the level of genes that reflects differences in the DNA sequences of different organisms.

In eukaryotic cells, recombination typically occurs during meiosis, a form of cell division that produces gametes, or egg and sperm cells. In the first phase of meiosis, the homologous pairs of maternal and paternal chromosomes align. During the alignment, the arms of the chromosomes can overlap and temporarily fuse, causing a crossover. Crossovers result in recombination and the exchange of genetic material between the maternal and paternal chromosomes. As a result, offspring can have different combinations of genes than either of their parents. Genes that are located farther apart on the same chromosome have a greater likelihood of undergoing recombination, which means they have a greater recombination frequency.

Scientific Ideas That Are Important to Think About During This Task

- Recombination is a process in which genes are broken apart and mixed into new combinations. This process takes place during meiosis.
- Most genes interact with many other genes, and some combinations of genes create better-adapted individuals than others.

Timeline

Approximately one class period.

Materials and Preparation

The items needed for this investigation are listed in Table 4.5. In preparation for this task, we recommend that you familiarize yourself with the online simulation and make sure it can be accessed on student devices.

Table 4.5. Required Materials for Task 24

Item	Quantity
Computer or tablet with internet connection	1 per group
Student Handout	1 per student

Safety Precautions

This task does not require any specific safety precautions.

Procedure

This lesson plan is only a suggestion. It is included here to illustrate how you can facilitate student thinking during this task. We encourage you to modify this lesson plan by asking different questions, using different examples, and providing different scaffolds as appropriate to better meet the needs of students in your class.

Putting Ideas on the Table (50 minutes)

1. Introduce the task using the PowerPoint template we have provided, which can be downloaded from the book's Extras page at *www.nsta.org/mbi-biology*.

2. Students should load the online simulation at *https://short.concord.org/lls* and begin at the heredity level.

3. As students work through the Student Handout and the various levels of the simulation, be sure to use Back Pocket Questions (see p. 346) to help them connect what they are seeing in their data.

4. After students have finished the simulation and answered the questions on the handout, conduct a whole-class discussion to review the results.

Adding Information to the Summary Table (10 minutes)

1. Give students 5 minutes to decide what to add to the Summary Table at the end of the Student Handout.

2. Have one student from each group share what the group figured out, how they know (their evidence for what they figured out), and how this information will help them explain the anchoring phenomenon.

3. Once each group has shared, ask the entire class to decide what should be added to each column of the class Summary Table for Task 24. Help students reach consensus about what to add to the Summary Table. Only add an idea to the Summary Table if everyone in the class agrees with that idea.

Back Pocket Questions

As students work in groups, it is important to engage with each group to help press and extend students' thinking around the ideas at play in this task. Following are some example questions you might ask:

1. Helping students get started: How would you describe what *genetic recombination* means? How is this similar to and different from mutations?

2. Pressing further: What patterns do you see in the simulation data? How does this relate to the mussel lures?

3. Following up: What makes you think that? Can you say more?

Filling Out the Summary Table

Table 4.6 includes examples of the responses students may come up with when they fill out the Summary Table. This is provided here only as a behind-the-scenes roadmap and is not meant to be shared with students.

Table 4.6. Example Summary Table for Task 24

What we learned	How it helps us explain the phenomenon
Recombination is a process in which genes are mixed into new combinations that result in variation of traits. Some combinations of genes create more fit, better-adapted individuals than others, which may lead to certain traits becoming more common in a population.	Recombination has led to a number of different variations in mussel lures. Some are more successful than others and thus are more common in the population.

Hints for Implementing This Task

- The Deer Mice simulation used here is quite complex. We suggest spending time using the simulation before introducing it to students so that you fully understand how it works and the specific concepts it links together through simulation.

- An online tutorial for the simulation is available at *https://connectedbio.org/resources/connectedbio-interactive-tutorial.mp4*.

- Additional materials to scaffold this simulation can be found at *https://learn.concord.org/cbio-deer-mouse*.

Possible Extensions

- The Deer Mice simulation is quite powerful, as it connects a number of different concepts in biology at three levels of scale: heredity, organism, and population. You may choose to explore the simulation further, making additional connections across biology to the ideas of heredity that this task focuses on.

- While the task is set up for students to explore the data, you may choose to conduct specific experiments within the simulation as a whole class.

Student Handout

TASK 24. GENETIC RECOMBINATION

Introduction

In the previous task, you looked at an important cause of variation of traits: mutation. Once the genes exist in a population through mutation, however, an even more important source of variation occurs: genetic recombination. Recombination is a process in which pieces of DNA are broken and recombined to produce new combinations of alleles. This recombination process creates diversity at the level of genes that reflects differences in the DNA sequences of different organisms.

Initial Ideas

Before beginning this task, take a few minutes to think about variation through mutation and genetic recombination. Describe how mutation can lead to variation.

How are mitosis and meiosis different?

Your Task

Your task is to explore variation caused by recombination in a population of deer mice using an online simulation.

1. Launch the ConnectedBio Deer Mice online simulation at *https://short.concord.org/lls*.

2. Once the simulation has loaded, be sure the red Heredity button on the left side of the screen is selected.

3. Choose a pair of mice to breed from one of the nests in the left pane, and complete the Punnett square on the following page by filling in the genotypes for the mother and father, using the notations in the key provided. You can see their individual genotypes by clicking on each and looking at the panel on the right. Then answer the question below the Punnett square.

Mother Genotype

Father Genotype

Key
- Light brown fur: (R^{L}R^L)
- Medium brown fur: (R^{L}R^D or R^{D}R^L)
- Dark brown fur: (R^{D}R^D)

a. What are the expected genotypes of the offspring of this pair?

2. Breed the parent mice by clicking on Breed. Once you do, you can see the resulting offspring of each litter, as well as the ratios in the total number of offspring for each parent pair. Be sure to explore both the left and right panes to see the options you have for visualizing the genotypes and phenotypes of the offspring, especially the graph function on the right. In the space below, create a table to collect data on the genotype of the breeding pair. Once you have bred the pair over multiple generations and explored this section, answer the following questions.

a. Why do some of the pie charts change as you breed the mice?

b. Why do the pie charts stabilize?

c. Where are the numbers different from the ratios in a Punnett square? Where are they the same?

4. Now click on the Organism button and read the instructions for comparing organisms of different genotypes at the cellular level. The goal here is to connect the organism's genotype to actual changes in function of the cells that produce hair color. Answer the following questions.

   a. How did the genotype of mice from your samples affect the cells that produce hair pigment?

b. At the cellular level, how does the structure of the receptor cause the difference between light brown and dark brown hair colors?

5. Finally, choose the Population button and read the instructions on the right. Here you can simulate different phenotypes for hair color and explore the impact of differences in this trait on the survivability and population dynamics of these organisms. In the space below, create a table to collect data helpful for answering the following question.

a. What happens to the population over time as you change various factors within the simulation?

6. Use what you learned to answer this wrap-up question. In this simulation, you explored deer mice at three levels: in terms of their heredity, the impact of their genotype at the cellular level, and the impact of their phenotype at the population level. How did these three levels affect one another in your simulation? Provide evidence for your response.

Some Useful Ideas From My Teacher

You can keep track of useful ideas from your teacher in the space below.

What We Figured Out

Now that you have completed this task, take a few minutes to fill out the Summary Table below with the other students in your group. This table will help you keep track of what you figured out during the task. You will then have an opportunity to share your ideas as we fill out our class Summary Table.

Summary Table

What we learned	How we know

How it helps us explain the phenomenon

Teacher Notes

Task 25. I Will Survive!

Purpose

In the previous tasks, students worked to understand the ways in which variation of traits and new adaptations occur through mutation and recombination. The goal of this task is for students to explain the importance of variation in evolution. In the simulation, students act as birds with different types of beaks competing for available food sources to see which will help the species survive in its environment. As the environment changes, students have to identify traits that would increase the species survival rate. Ultimately, students discover that variation is essential to the survival of species, as those species that are best adapted to their environment will survive and reproduce at higher rates.

Important Life Science Content

Adaptations are mutations, or genetic changes, that help an organism survive in its environment. If a mutation is helpful, it is passed down from one generation to the next. As more and more organisms inherit the mutation, it becomes a typical trait of the species and has become an adaptation. An adaptation can be structural, meaning it is a physical part of the organism, or behavioral, affecting the way an organism acts. Adaptations usually develop in response to a change in an organism's habitat. For example, changes in food availability in the Galapagos Islands led to new adaptations in finch beaks, an example students will explore in the next two tasks.

Scientific Ideas That Are Important to Think About During This Task

- Adaptations are variations in traits that help organisms survive in their environment.
- The environment determines which variation of a trait will help an organism survive.
- A population has genetic variations, possibly due to mutations. Favorable variations may allow an organism to be better adapted to its environment and survive to reproduce.

Timeline

Approximately one class period.

Materials and Preparation

The items needed for this investigation are listed in Table 4.7. Each group will need an assortment of "foods" (pennies, marbles, rice, beans, and marshmallows) and "beaks" (tongs, test tube holders, binder clips, and toothpicks). Other materials can be added or substituted as available.

Table 4.7. Required Materials for Task 25

Item	Quantity
Pennies	Several per group
Marbles	Several per group
Grains of rice	Several per group
Beans	Several per group
Marshmallows	Several per group
Tongs	Several per group
Test tube holders	Several per group
Binder clips	Several per group
Toothpicks	Several per group
Plastic cup	1 per student
Student Handout	1 per student

Safety Precautions

Remind students to follow all normal safety rules. In addition, tell students to take the following safety precautions:

- Never taste or eat any food items used in a lab activity.
- Clean up any materials that fall on the floor immediately to avoid a slip or fall hazard.
- Wash hands with soap and water when the activity is completed.

Procedure

This lesson plan is only a suggestion. It is included here to illustrate how you can facilitate student thinking during this task. We encourage you to modify this lesson plan by asking different questions, using different examples, and providing different scaffolds as appropriate to better meet the needs of students in your class.

Putting Ideas on the Table (50 minutes)

1. Introduce the task using the PowerPoint template we have provided, which can be downloaded from the book's Extras page at *www.nsta.org/mbi-biology*.
2. Students should each choose a "beak" (multiple students can have the same beak if necessary) and run the simulation twice, keeping careful track of the data throughout.
3. Once the trials are completed, students should work together in groups to graph and analyze the data.

Adding Information to the Summary Table (10 minutes)

1. Give students 5 minutes to decide what to add to the Summary Table at the end of the Student Handout.

2. Have one student from each group share what the group figured out, how they know (their evidence for what they figured out), and how this information will help them explain the anchoring phenomenon.

3. Once each group has shared, ask the entire class to decide what should be added to each column of the class Summary Table for Task 25. Help students reach consensus about what to add to the Summary Table. Only add an idea to the Summary Table if everyone in the class agrees with that idea.

Back Pocket Questions

As students work in groups, it is important to engage with each group to help press and extend students' thinking around the ideas at play in this task. Following are some example questions you might ask:

1. Helping students get started: What do the tools represent in this simulation? What do the other items, such as the pennies and beans, represent?

2. Pressing further: What are you noticing in your data? Do you see any patterns in which types of beaks work the best?

3. Following up: What makes you think that? Can you say more?

Filling Out the Summary Table

Table 4.8 includes examples of the responses students may come up with when they fill out the Summary Table. This is provided here only as a behind-the-scenes roadmap and is not meant to be shared with students.

Table 4.8. Example Summary Table for Task 25

What we learned	How it helps us explain the phenomenon
Variation leads to different survival rates among individuals. Traits that positively affect survival are more likely to be reproduced and thus become more common in the population.	There is variation in *Lampsilis* mussel lure morphology. Mussels with lures that best mimic the prey species of a host fish are more likely to survive and reproduce, and thus these lures become more common in the population.

Possible Extensions

- Groups could conduct short investigations into other animal adaptations to relate back to the finch beaks.

Student Handout

TASK 25. I WILL SURVIVE!

Introduction

In the previous tasks, you explored the causes of variations. You learned that new traits occur because of mutations in DNA, and then recombination of traits within a population leads to further variation. When a trait is useful and allows organisms of a species to survive better, it is referred to as an adaptation. In this task, you will further examine the impact of variations on a population and the important role the environment plays in this process.

Initial Ideas

Before beginning this task, take a few minutes to think about variations within a population. What causes new variations of a trait to occur?

Once a new trait occurs, what causes it to vary even more within a population?

Your Task

Your task is to explore the impact of variations on a population and the important role the environment plays by simulating competition for resources among birds in a given ecosystem, using the available materials to represent beaks and food items.

Available Materials

You and your group may use the following materials during this task:

- Pennies
- Marbles
- Rice
- Beans
- Marshmallows

- Tongs
- Test tube holders
- Binder clips
- Toothpicks
- Plastic cup

Safety Precautions

Follow all normal lab safety rules. In addition, be sure to take the following safety precautions:

- Never taste or eat any food items used in a lab activity.
- Clean up any materials that fall on the floor immediately to avoid a slip or fall hazard.
- Wash hands with soap and water when the activity is completed.

Procedure

1. Working in your group, you will simulate competition for resources among birds in a given ecosystem. Each student in the group should choose one of the tools that represent beaks to pick up the food items.
2. Using your "beak," pick up as many food items as you can in each round and put them into the cup.
3. Run the simulation twice, keeping careful track of the data throughout.
4. Once both trials are completed, work together with your group to graph and analyze the data.

Trial 1

Beak Type	Round								Total	Food (prey)
	1	2	3	4	5	6	7	8		
Tongs										Pennies
										Marbles
										Rice
										Beans
										Marshmallows
Test tube holder										Pennies
										Marbles
										Rice
										Beans
										Marshmallows
Binder clip										Pennies
										Marbles
										Rice
										Beans
										Marshmallows
Toothpick										Pennies
										Marbles
										Rice
										Beans
										Marshmallows
Two toothpicks										Pennies
										Marbles
										Rice
										Beans
										Marshmallows

Trial 2

Beak Type	Round								Total	Food (prey)
	1	2	3	4	5	6	7	8		
Tongs										Pennies
										Marbles
										Rice
										Beans
										Marshmallows
Test tube holder										Pennies
										Marbles
										Rice
										Beans
										Marshmallows
Binder clip										Pennies
										Marbles
										Rice
										Beans
										Marshmallows
Toothpick										Pennies
										Marbles
										Rice
										Beans
										Marshmallows
Two toothpicks										Pennies
										Marbles
										Rice
										Beans
										Marshmallows

Key

More than 6 foods (GREEN) means the bird survives.

Between 3 and 6 foods (YELLOW) means the bird survives for 1 more round.

Fewer than 3 foods (RED) means the bird does not survive.

Graph

Create a cluster bar graph with the food types on the x-axis and the total number of foods collected on the y-axis. Graph BOTH trials on the same graph. If a beak type participated in both trials, graph the AVERAGE total. Be sure to make a legend to indicate the colors used for each beak type (see the Key).

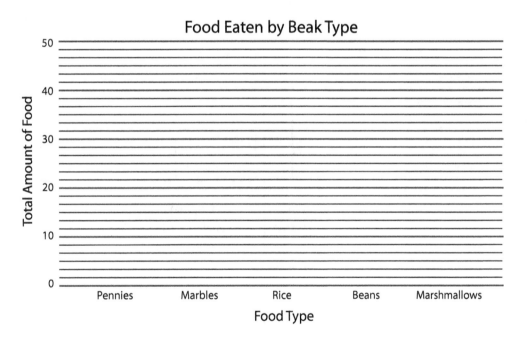

Legend

Tongs =	Test tube holder =	Binder clip =	Toothpick =	Two toothpicks =

Analysis

1. Explain the effect(s) the mutations had on the survivability of the affected birds. Use data from the lab to support your answer.

2. Which bird(s) seemed best adapted to picking up each of the following?

Pennies	
Marbles	
Rice	
Beans	
Marshmallows	

3. Which bird(s) survived to the end? Use data from the lab to explain WHY they survived.

4. Which bird(s) seemed least adapted to picking up each of the following?

Pennies	
Marbles	
Rice	
Beans	
Marshmallows	

5. Which bird(s) died off first? Use data from the lab to explain WHY they died off first.

6. In your own words, explain what an adaptation is.

Some Useful Ideas From My Teacher

You can keep track of useful ideas from your teacher in the space below.

What We Figured Out

Now that you have completed this task, take a few minutes to fill out the Summary Table below with the other students in your group. This table will help you keep track of what you figured out during the task. You will then have an opportunity to share your ideas as we fill out our class Summary Table.

Summary Table

What we learned	How we know

How it helps us explain the phenomenon

Midunit Model Revision for Unit 4

Purpose

Now that your class is roughly halfway through the unit, it is important for students to go back to the initial models constructed on the first day and revise them based on what they have learned so far. Students may choose between two different strategies for model revisions. After negotiating what should be added, revised, or removed from their models, the groups may do either of the following:

- Redraw their models.
- Use sticky notes to keep track of their revisions for the final model revision near the end of the unit.

They should make the choice between these options based on the amount of time available for the model revision process. It is important, however, that students make decisions about revisions based on what fits with the evidence from the tasks or is consistent with the scientific ideas at play. We have provided a model revision PowerPoint template for your use, which can be downloaded from the book's Extras page at *www.nsta.org/mbi-biology*, and Figure 4.7 contains resources to help with modeling.

Figure 4.7. Resources to Help With Modeling

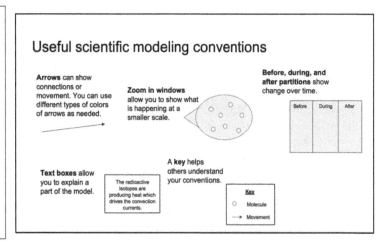

Although it does take time, we recommend a share-out session after the model revision. This allows for new ideas to be shared across the whole class, as well as discussion about what should and should not be included in the models. Have the members of each group stand up and describe how and why they changed their model. If time is limited, you could ask each group to focus on just one change. Provide time for questions from the class, and be sure to ask clarifying questions while comparing and contrasting the ideas presented across the groups.

Teacher Notes

Task 26. Galapagos Finches

Purpose

In the first task in this unit, students identified the relationship between host fish diversity and *Lampsilis* mussel lure diversity: Host fish are attracted to different prey species, and the mussels mimic this prey with their lifelike lures, thereby giving rise to an increase in mussel lure diversity. This task presents students with another classic example of evolution by natural selection: the work of Peter and Rosemary Grant, who observed and measured natural selection by studying the finch species on the Galapagos Islands in the Pacific Ocean off South America. Students learn about the Grants' experimental design, their fieldwork, and environmental changes of 1977 and 1983 that led to their important findings. In addition, students are presented with evidence to demonstrate how it is possible that one species of finch arrived on the Galapagos Islands and subsequently evolved into the 13 species of finches that inhabit the islands today. Similarly, one species of *Lampsilis* mussel may have diversified into the many different species that we see today. Working in groups, students come up with an explanation for how it is possible that one species of *Lampsilis* mussel evolved into many different species.

Important Life Science Content

Charles Darwin proposed the theory of evolution by natural selection in his most important work, *On the Origin of Species*, in 1859, many years after his voyage on HMS *Beagle*. Evolution by natural selection was such a radical idea at the time that he presented his idea using artificial selection as a model for natural selection. Darwin was never able to provide empirical evidence to support his idea that the mechanism of evolution was natural selection. However, his work inspired the next generation of scientists, and more than 100 years later, Peter and Rosemary Grant were able to successfully do what Darwin could not: observe and measure natural selection.

Scientific Ideas That Are Important to Think About During This Task

- Evolution is change in the heritable characteristics of biological populations over successive generations.
- Natural selection is the mechanism by which some organisms are better adapted to their environments and produce more offspring.

Timeline

Approximately one class period.

Materials and Preparation

The items needed for this investigation are listed in Table 4.9. In preparation for this task, we recommend that you familiarize yourself with the online video and how to show it to the entire class.

Table 4.9. Required Materials for Task 26

Item	Quantity
Computer or projector to show the video	1 per class
Student Handout	1 per student

Safety Precautions

This task does not require any specific safety precautions.

Procedure

This lesson plan is only a suggestion. It is included here to illustrate how you can facilitate student thinking during this task. We encourage you to modify this lesson plan by asking different questions, using different examples, and providing different scaffolds as appropriate to better meet the needs of students in your class.

Putting Ideas on the Table (50 minutes)

1. Review model revisions and Summary Tables from the previous tasks in this unit, focusing on how they apply to the anchoring phenomenon.

2. Highlight the Lure Morphology task, driving home the relationship between host fish diversity and *Lampsilis* mussel diversity. Perhaps pose the question, "Is it possible one species of *Lampsilis* mussel changed into many different species of freshwater mussels?"

3. Introduce Charles Darwin and Peter and Rosemary Grant. Following this introduction, students should read the summary introducing Darwin and Peter and Rosemary Grant included on the Student Handout.

4. Distribute the questions about the video and have students read through the instructions.

5. Play the HHMI BioInteractive video "The Origin of Species: The Beak of the Finch" at *www.biointeractive.org/classroom-resources/origin-species-beak-finch*. Tell students to answer the first nine questions on the handout as the video plays. These questions are designed to guide students to answer the final question, explaining and modeling how it's possible that one species of *Lampsilis* mussel evolved into many different species. Stop the video frequently to review answers to the questions and discuss important points.

6. Give groups time to meet and compare answers. Then ask them to answer the last question on the handout as a group.

7. Have students share their answers to the last question and discuss as a class.

Adding Information to the Summary Table (10 minutes)

1. Give students 5 minutes to decide what to add to the Summary Table at the end of the Student Handout.

2. Have one student from each group share what the group figured out, how they know (their evidence for what they figured out), and how this information will help them explain the anchoring phenomenon.

3. Once each group has shared, ask the entire class to decide what should be added to each column of the class Summary Table for Task 26. Help students reach consensus about what to add to the Summary Table. Only add an idea to the Summary Table if everyone in the class agrees with that idea.

Back Pocket Questions

As students work in groups, it is important to engage with each group to help press and extend students' thinking around the ideas at play in this task. Following are some example questions you might ask:

1. Helping students get started: What are the Grants trying to figure out? What data are they collecting and why?

2. Pressing further: How do the results from the Grants' work relate to the *Lampsilis* mussel? What patterns did they notice in their finch data?

3. Following up: What makes you think that? Can you say more?

Filling Out the Summary Table

Table 4.10 includes examples of the responses students may come up with when they fill out the Summary Table. This is provided here only as a behind-the-scenes roadmap and is not meant to be shared with students.

Table 4.10. Example Summary Table for Task 26

What we learned	How it helps us explain the phenomenon
Evolution is the change in characteristics over time. Changes in the environment can cause large changes in populations, even resulting in new species.	One species of *Lampsilis* mussel diversified into many different species by natural selection.

Hints for Implementing This Task

- We suggest periodically stopping the video to provide time for students to answer the questions and have discussions to call out important points.

Possible Extensions

- Create three bell curves to illustrate directional selection when discussing questions 5 and 6, using the time periods 1973 (pre-drought), 1978 (post-drought), and 1984 (following a strong El Niño). Note that the environment (food type/availability) was favoring or selecting for a particular phenotype (beak size) during each of these time periods.

- An educator's guide to the video with additional extensions is available at *www.biointeractive. org/sites/default/files/BeakofFinch–Educator-film.pdf*.

Student Handout

TASK 26. GALAPAGOS FINCHES

Introduction

In 1831, at the age of 22, Charles Darwin set sail on HMS *Beagle* as a member of a research team whose goal was to survey the coast of South America. As a naturalist, Darwin observed and collected many different specimens of plants, rocks, animals, and fossils over the six-year journey. Darwin is perhaps most famous for the observations he made and specimens he collected on the Galapagos Islands, which included iguanas, tortoises, plants, and many different species of birds. While on the islands, Darwin believed he had collected different kinds of birds, including finches, wrens, and warblers. However, after returning to England, while classifying these bird specimens, he and other scientists discovered that the birds had been misidentified in the field and were in fact various species of finches. Each species had variations in terms of beak size, plumage, and body size. To explain these findings, Darwin proposed that these slight variations were the result of evolution by natural selection, the process whereby organisms better adapted to their environment tend to survive and produce more offspring. Natural selection is the mechanism of evolution.

Darwin never went back to the Galapagos Islands to test his theory of evolution but inspired many with his radical ideas. It was not until 1973, when Princeton University scientists Peter and Rosemary Grant began their research on the finches on the Galapagos Islands, that someone first measured and observed the process of natural selection in action. You will watch a video that tells the story of how the Grants were able to empirically show what Darwin had proposed over a century earlier: that the 13 species of finches on the Galapagos Islands all diversified from one ancestral population and that new species come from preexisting species.

Initial Ideas

Before beginning this task, take a few minutes to think about the Grants' research on natural selection. How do you think Peter and Rosemary Grant measured natural selection, a process that happens over long periods of time?

Your Task

Answer the following questions as you watch the video "The Origin of Species: The Beak of the Finch."

1. How did the Galapagos Islands form? Was there any life on these islands at the time of their formation?

2. What differences exist between each of these volcanic islands? (*Hint:* habitats)

3. What important differences exist among the 13 species of finches?

4. How did the Galapagos end up with so many species of finches? (List two possibilities.)

5. What happened in 1977? How did this affect the medium ground finch (MGF) population?

6. What happened in 1983? How did this affect the medium ground finch (MGF) population?

7. Define *species*.

8. What two factors keep different species from mating?

9. Based on what you learned from the video, explain how one species of finch diversified into 13 distinct species.

10. Work in your groups to answer this last question: Based on what you have learned about how species can change over time, describe how it's possible for one species of *Lampsilis* mussel to evolve into 10 different species of *Lampsilis* mussels. Use the terms *variation*, *selection*, *inheritance*, and *time* in your answer.

Some Useful Ideas From My Teacher

You can keep track of useful ideas from your teacher in the space below.

What We Figured Out

Now that you have completed this task, take a few minutes to fill out the Summary Table below with the other students in your group. This table will help you keep track of what you figured out during the task. You will then have an opportunity to share your ideas as we fill out our class Summary Table.

Summary Table

What we learned	How we know

How it helps us explain the phenomenon

Teacher Notes

Task 27. Does Beak Size Matter?

Purpose

In the previous task, students learned about how Charles Darwin and Peter and Rosemary Grant contributed to our understanding of evolution by natural selection. Ultimately, their work helped explain how one ancestral species of finch could evolve into 13 species. The goal of this task is to use data collected by Peter and Rosemary Grant as empirical evidence to illustrate natural selection. Students produce line graphs showing food type/abundance and beak depth changes over time. In the next task, after interpreting these graphs, students will create an investigation similar to the Grants' in which they measure natural selection in *Lampsilis* mussels.

Important Life Science Content

Natural selection is the process through which populations of living organisms adapt and change. Individuals in a population are naturally variable, meaning they are all different in some ways. This variation means that some individuals have traits better suited to the environment than others. Individuals with adaptive traits, or adaptations, are more likely to survive and reproduce. These individuals then pass the adaptive traits on to their offspring. Over time, these advantageous traits become more common in the population. Through this process of natural selection, favorable traits are transmitted through generations. Over a long enough period of time and in the right conditions, natural selection can lead to speciation, where one species gives rise to a new and distinctly different species. It is one of the processes that drive evolution and help explain the diversity of life on Earth.

Scientific Ideas That Are Important to Think About During This Task

- Natural selection is the process whereby organisms better adapted to their environment tend to survive and produce more offspring. Natural selection is the mechanism of evolution.

- Speciation is the process by which new species evolve.

Timeline

Approximately one to two class periods.

Materials and Preparation

The items needed for this investigation are listed in Table 4.11.

Table 4.11. Required Materials for Task 27

Item	Quantity
Graph paper	1 sheet per group
Student Handout	1 per student

Safety Precautions

This task does not require any specific safety precautions.

Procedure

This lesson plan is only a suggestion. It is included here to illustrate how you can facilitate student thinking during this task. We encourage you to modify this lesson plan by asking different questions, using different examples, and providing different scaffolds as appropriate to better meet the needs of students in your class.

Putting Ideas on the Table (50 minutes)

1. Review the video "The Origin of Species: The Beak of the Finch" from the previous task. Emphasize the Grants' experimental design and data collection techniques. These datasets were important in explaining how one species of finch diversified into 13 species. In this task, students will use data collected by Peter and Rosemary Grant to explain natural selection.

2. Have students follow the instructions on their handouts to produce two line graphs, showing food type/abundance over time and beak depth over time, then interpret these graphs by answering the series of questions on the handout.

3. Review and discuss these questions as a whole class.

Adding Information to the Summary Table (10 minutes)

1. Give students 5 minutes to decide what to add to the Summary Table at the end of the Student Handout.

2. Have one student from each group share what the group figured out, how they know (their evidence for what they figured out), and how this information will help them explain the anchoring phenomenon.

3. Once each group has shared, ask the entire class to decide what should be added to each column of the class Summary Table for Task 27. Help students reach consensus about what to add to the Summary Table. Only add an idea to the Summary Table if everyone in the class agrees with that idea.

Back Pocket Questions

As students work in groups, it is important to engage with each group to help press and extend students' thinking around the ideas at play in this task. Following are some example questions you might ask:

1. Helping students get started: What patterns in the data do you notice? Why did the Grants measure the number of small and large seeds?

2. Pressing further: What would have happened in the data if the events of 1977 and 1983 had not occurred?

3. Following up: What makes you think that? Can you say more?

Filling Out the Summary Table

Table 4.12 includes examples of the responses students may come up with when they fill out the Summary Table. This is provided here only as a behind-the-scenes roadmap and is not meant to be shared with students.

Table 4.12. Example Summary Table for Task 27

What we learned	How it helps us explain the phenomenon
Natural selection is the process whereby organisms better adapted to their environment tend to survive and produce more offspring. This can lead to the creation of new species, known as speciation.	As the environment around the mussels changed, natural selection acted on the variations of mussel lures..

Hints for Implementing This Task

- Students may require additional instruction or a review on graphing to complete this task.

Possible Extension

- Students may explore the lab on bird beaks as tools at *www.biointeractive.org/classroom-resources/beaks-tools-selective-advantage-changing-environments*.

Student Handout

TASK 27. DOES BEAK SIZE MATTER?

Introduction

From 1977 to 1978, the island of Daphne Major went 500 days without precipitation. As a result, the number and type of seeds produced by plants changed. Birds, specifically the numerous species of finches on the island, depended on these seeds for food. Finches use their beaks to crack open seeds and consume the fruit inside. Variation in beak size and depth allows certain species of finch to be more successful at cracking open seeds of different sizes. For example, large beaks are better at cracking open large, tough seeds. Think of the beak like a tool; the right tool is needed for the job. In this task, you will use data collected by scientists Peter and Rosemary Grant to better understand the relationship between seed size availability and the average size of finch beaks.

Initial Ideas

Before beginning this task, take a few minutes to think about the variation observed in finch beaks. Based on what you learned from the video, describe the variation seen in the finch beaks on Daphne Major.

Your Task

Your task is to use data collected by scientists Peter and Rosemary Grant to create line graphs and answer questions to explore the relationship between the changing seed supply and the average size of finch beaks. This will help you better understand variation in finch beak size and depth.

Seed Supply

Use the data in Table SH27.1 to create a line graph showing how the total number of small and large seeds changed annually from 1976 to 1987. Then answer the questions that follow the graph.

Table SH27.1. Daphne Major Seed Data, 1976–87

Year	Small seeds	Large seeds
1976	700	850
1977	100	900
1978	250	500
1979	270	970
1980	50	1,300
1981	200	1,200
1982	25	50
1983	300	100
1984	1,800	300
1985	2,200	300
1986	1,600	250
1987	3,050	600

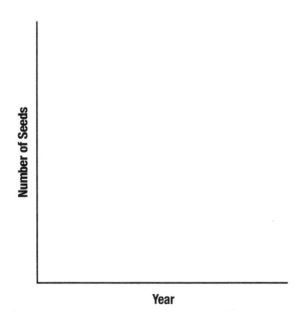

1. What years had the highest and lowest numbers of large seeds? How can you account for this?

2. What years had the highest and lowest numbers of small seeds? How can you account for this?

Beak Depth

Use the data in Table SH27.2 to create a line graph showing how the average finch beak depth changed from 1976 to 1987. Then answer the questions that follow the graph.

Table SH27.2. Daphne Major Finch Beak Depth Data, 1976–87

Year	Beak depth (mm)
1976	9.23
1977	9.35
1978	9.74
1979	9.78
1980	9.81
1981	9.75
1982	9.8
1983	9.71
1984	9.66
1985	9.62
1986	9.48
1987	9.31

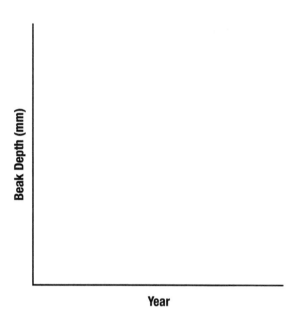

3. What years had the highest and lowest average finch beak depths? How can you account for this?

4. Compare the graphs side by side. Describe the relationship between seed size availability and beak depth.

5. In 1978, following a period of over 500 days with no precipitation, scientists returned to Daphne Major and discovered that finches with bigger beaks had survived at a higher rate than those with smaller beaks. Why?

6. Beak size is a variation passed from parent to offspring. In 1978, when scientists measured the beak size of the new generation of finches, they discovered there were many more birds with larger beaks. What happened?

7. From 1983 to 1985, the number of small seeds increased dramatically. What might have caused this change in seed supply? (*Hint:* Think back to the video.) How did this affect the size of bird beaks during this time period?

8. Beak size is a variation passed from parent to offspring. Based on what you learned in this task, what environmental factor/selective pressure was driving natural selection?

9. Similarly, a lifelike lure is a variation passed from parent to offspring in *Lampsilis* mussels. What environmental factor/selective pressure is driving natural selection in this case?

Some Useful Ideas From My Teacher

You can keep track of useful ideas from your teacher in the space below.

What We Figured Out

Now that you have completed this task, take a few minutes to fill out the Summary Table below with the other students in your group. This table will help you keep track of what you figured out during the task. You will then have an opportunity to share your ideas as we fill out our class Summary Table.

Summary Table

What we learned	How we know

How it helps us explain the phenomenon

Teacher Notes

Task 28. Where'd Those Mussels Come From?

Purpose

Students build on the previous task in which they learned that variation leads to different survival rates among individuals within a species. Mussels with variations of lures that best mimic the prey species of a host fish are more likely to survive and reproduce, and thus become more common in the population. In this task, students learn how an increase in selective pressures (host fish diversity) creates more opportunities for the development of new mussel species. They then design an investigation to measure natural selection in the mussel population. The goal of this task is to reinforce the concept that it is the environment (in this case the host fish) that selects for or determines whether a trait is an adaptation. These selective pressures drive natural selection.

Important Life Science Content

Speciation is how a new kind of plant or animal species is created. It occurs when a group within a species separates from other members of its species and develops its own unique characteristics. The demands of a different environment or the characteristics of the members of the new group will differentiate the new species from its ancestors. As seen in previous tasks, the Galapagos finches are an example. Different finch species live on different islands in the Galapagos archipelago. The finches are isolated from one another by the ocean. Over millions of years, each species of finch developed a unique beak that is especially adapted to the kinds of food it eats. Some finches have large, blunt beaks that can crack the hard shells of nuts and seeds. Other finches have long, thin beaks that allow them to probe into cactus flowers without being poked by the cactus spines. Still other finches have medium-size beaks that can catch and grasp insects. Because they are isolated, the birds on each island don't breed with those on the other islands and have therefore developed into unique species with unique characteristics.

Speciation can occur because of a number of different situations. Most commonly, a physical barrier, such as a mountain range or waterway, makes it impossible for populations to breed with one another. Each species develops differently based on the demands of its unique habitat or the genetic characteristics of the group that are passed on to offspring. The Galapagos finches are an example. Populations can also be spread out over such a large geographic area that it is not possible for members of the species to mate with other members outside of their own local region. Instead of being separated by a physical barrier, the species are separated by differences in the same environment.

Scientific Ideas That Are Important to Think About During This Task

- Speciation is the process by which new species evolve.
- Speciation most often occurs because of a physical barrier that impedes the genetic flow of traits between populations.

Timeline

Approximately one class period.

Materials and Preparation

The items needed for this investigation are listed in Table 4.13.

Table 4.13. Required Materials for Task 28

Item	Quantity
Whiteboard or poster paper	1 sheet per group
Student Handout	1 per student

Safety Precautions

This task does not require any specific safety precautions.

Procedure

This lesson plan is only a suggestion. It is included here to illustrate how you can facilitate student thinking during this task. We encourage you to modify this lesson plan by asking different questions, using different examples, and providing different scaffolds as appropriate to better meet the needs of students in your class.

Part 1 (20 minutes)

1. Introduce Part 1 of the task using the PowerPoint template we have provided, which can be downloaded from the book's Extras page at *www.nsta.org/mbi-biology*. Discuss the many Virginia watersheds, highlighting the Clinch River and the New River.

2. Students should begin working on Part 1 of the task in groups using their handouts, which include a data table illustrating the number of native fish species and mussel species from Virginia, the Clinch River, and the New River. They are asked to create a bar graph to display these data and then use this information to come up with an explanation as to why there are so many different species of mussels.

3. Once students have completed Part 1, discuss the answers to the questions as a whole class, focusing on the idea of speciation.

Part 2 (30 minutes)

1. Introduce Part 2 of the task, explaining that students will use information from this and the previous few tasks to design an investigation that would allow them to observe and measure natural selection in freshwater mussels.

2. Review the Grants' experimental design with the class:

   - The island of Daphne Major was an ideal study site because the finches had little competition (the only other finch on the island is the cactus finch) and few predators. Therefore, the selective pressure controlling the survival of the medium ground finch is the weather and subsequent availability of food.
   - The Grants used mist netting to capture finches in order to measure beak size and bird weight, collect DNA samples, and band the birds.
   - They conducted fieldwork annually to observe changes.

3. Explain that in designing their investigation, students must consider the following:

   - What would an ideal study site look like in order to observe and measure changes in *Lampsilis* mussel lure morphology?
   - What is the selective pressure influencing the mussels' ability to reproduce?
   - What data would be important to collect? How would you collect these data?
   - How often should these measurements be taken?

4. After designing their investigations, have students present them to the class.

Adding Information to the Summary Table (10 minutes)

1. Give students 5 minutes to decide what to add to the Summary Table at the end of the Student Handout.

2. Have one student from each group share what the group figured out, how they know (their evidence for what they figured out), and how this information will help them explain the anchoring phenomenon.

3. Once each group has shared, ask the entire class to decide what should be added to each column of the class Summary Table for Task 28. Help students reach consensus about what to add to the Summary Table. Only add an idea to the Summary Table if everyone in the class agrees with that idea.

Back Pocket Questions

As students work in groups, it is important to engage with each group to help press and extend students' thinking around the ideas at play in this task. Following are some example questions you might ask:

1. Helping students get started: Why are there so many different species of mussels in this river system?

2. Pressing further: How does the change in predators lead to new species of mussels? What is happening?

3. Following up: What makes you think that? Can you say more?

Filling Out the Summary Table

Table 4.14 includes examples of the responses students may come up with when they fill out the Summary Table. This is provided here only as a behind-the-scenes roadmap and is not meant to be shared with students.

Table 4.14. Example Summary Table for Task 28

What we learned	How it helps us explain the phenomenon
Speciation most often occurs because of a physical barrier that impedes the genetic flow of traits between populations.	Host fish are attracted to different prey. *Lampsilis* mussels mimic this prey with their lifelike lures, thereby giving rise to an increase in lure diversity.

Hints for Implementing This Task

- Students often have trouble making the connection to speciation. A discussion of mechanisms of speciation with further examples may be appropriate here.

Possible Extensions

- For additional engagement with the idea of speciation, students can explore HHMI Bio-Interactive's Lizard Evolution Virtual Lab at *www.biointeractive.org/classroom-resources/lizard-evolution-virtual-lab*.

Student Handout

TASK 28. WHERE'D THOSE MUSSELS COME FROM?

Introduction

Speciation is the process by which new species arise. You have seen an example of this with Darwin's finches in the Galapagos Islands. There are a number of ways this can occur. Often, some sort of physical barrier (for example, a mountain range, canyon, or islands) will separate populations of a species, which will adapt to the new environments through natural selection and, over time, become distinct species.

Initial Ideas

Before beginning this task, take a few minutes to think about speciation. How are new species created? Give an example using the Galapagos finches.

Your Task

Your task in Part 1 is to look at data on mussel species in Virginia rivers for evidence of what may have caused the diversity of species seen today. Then, in Part 2, you will design an investigation to show natural selection at work on the *Lampsilis* mussels. Unfortunately, this is not an investigation you will be able to conduct (the Grants spent decades on their research!), but you can learn what your peers think will provide good empirical support for natural selection.

Part 1: Examine Data

The freshwater mussels in Virginia are among the most diverse in the United States, with 77 different species. Within Virginia, mussel diversity varies widely among river systems (Figure SH28.1). The Clinch River has the highest diversity of mussels in the state, with 53 species, while the New River, just east of the Clinch, has low mussel diversity, with only 12 different species. In Part 1 of this task, you will determine the mechanism driving mussel diversity in Virginia.

Figure SH28.1. Major River Drainages of Virginia

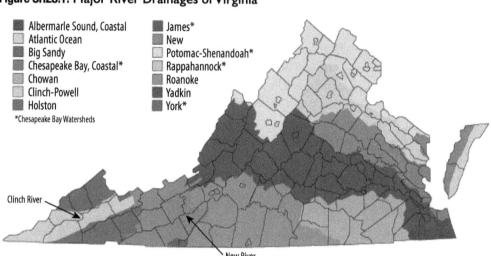

Albermarle Sound, Coastal
Atlantic Ocean
Big Sandy
Chesapeake Bay, Coastal*
Chowan
Clinch-Powell
Holston
*Chesapeake Bay Watersheds

James*
New
Potomac-Shenandoah*
Rappahannock*
Roanoke
Yadkin
York*

Clinch River

New River

1. Using the data presented in Table SH28.1, create a bar graph comparing the number of native fish species and the number of freshwater mussel species from the three locations.

Table SH28.1. Native Fish and Mussel Species Data Table

Location	Number of native fish species	Number of freshwater mussel species
Virginia	210	77
Clinch River	120	53
New River	46	12

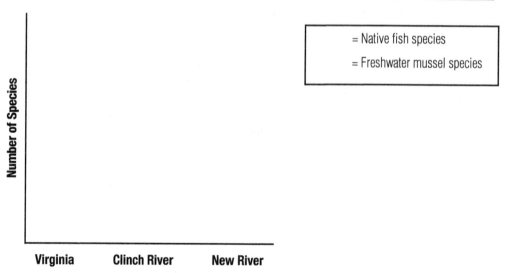

= Native fish species

= Freshwater mussel species

Number of Species

Virginia   Clinch River   New River

2. What is the relationship between the number of native fish species and the number of freshwater mussel species? How can you explain this pattern?

3. What do you think the native fish eat? Do you think any of them are fooled by the life-like lures of *Lampsilis* mussels?

4. How might a high number of native fish species (high diversity) increase the number of freshwater mussel species?

5. What would happen if the same assemblage of *Lampsilis* mussels were placed in two different river systems, where the host fish preyed on the red pucker-lip fish in one system and the yellow smiler fish in the other?

6. If, after a million years, the mussels with red and yellow lures from these two systems were reintroduced into the same river system, do you think they could successfully reproduce together?

Part 2: Design an Investigation

The Grants were able to observe and measure natural selection take place by studying medium ground finches on the island of Daphne Major. Every year since 1973, the Grants have returned to the island to gather data on weather, food type and abundance, and beak size. Through their careful work, they were able to document natural selection and provide empirical evidence to support Darwin's theory of evolution.

Using what you've learned about the Grants' work, design an investigation that would allow you to observe and measure natural selection in freshwater mussels. What data would you need to collect? How would you go about this data collection process? Create an investigation by answering the questions in the spaces provided.

Your Plan

What is your question?

What data will you collect?

**How will you
collect your
data?**

**What safety
precautions will
you follow?**

**How will you
analyze your
data?**

Some Useful Ideas From My Teacher

You can keep track of useful ideas from your teacher in the space below.

What We Figured Out

Now that you have completed this task, take a few minutes to fill out the Summary Table that follows with the other students in your group. This table will help you keep track of what you figured out during the task. You will then have an opportunity to share your ideas as we fill out our class Summary Table.

Summary Table

What we learned	How we know

How it helps us explain the phenomenon

Building Consensus Stage Summary

In the building consensus stage of an MBI unit (Figure 4.8), students finalize their group models, compare and contrast those models as a class as they work to reach a consensus, and co-construct a Gotta Have Checklist of the ideas and evidence that should be part of their final evidence-based explanations. Before beginning this stage, review the relevant sections in Chapters 1 and 2 for specifics. See Figures 4.3 and 4.4 (pp. 325–326) for examples of a Gotta Have Checklist and a final model for this unit. We have provided a PowerPoint template for this stage, which can be downloaded from the book's Extras page at *www.nsta.org/mbi-biology*.

Figure 4.8. Building Consensus Stage Summary

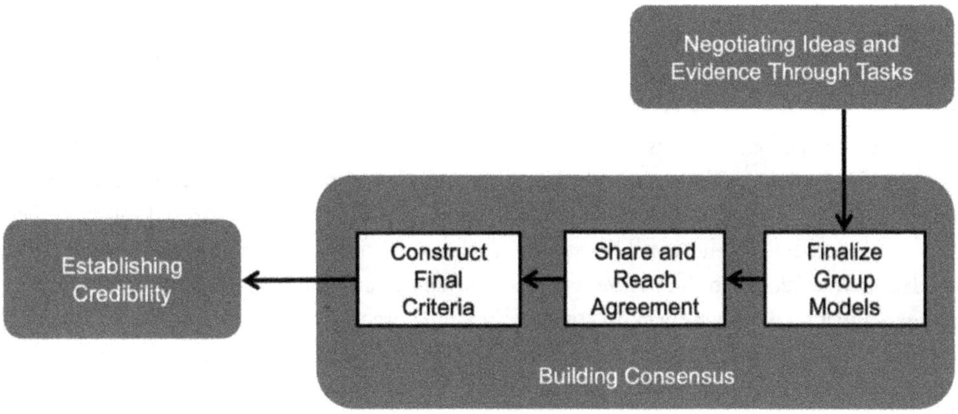

Finalize Group Models

Finalizing the models requires groups to review their previous models, decide what needs to be revised based on new ideas and understandings from the last set of tasks, and redraw the models so they can be more easily shared and used to build consensus in a whole-class setting. This usually takes about 30 minutes. We often scaffold this process by asking students to review the completed Summary Table and talk about what should be added to, removed from, or changed in their previous models. They should think about what new ideas can help them move toward a truly causal model that includes not only *what* happened but also *why* it happened. This requires that the mechanism at play be visible and well explained. We push the groups to make sure that items in their models are labeled, that any unseen components are made visible, and that the important ideas that surfaced throughout the unit are explicitly used in the models.

Share and Reach Agreement

We think a share-out session is needed here so the groups can learn from each other and begin to build consensus across the models. There are a number of ways to do this, including gallery walks. As the goal is to build consensus as a whole class, we usually opt to facilitate a share-out session in which each group comes to the front of the class, displays its model, and talks it through. We prompt students to ask questions, and the class works hard to compare and contrast ideas across the models.

Construct a List of Final Criteria

The goal of our last public record, the Gotta Have Checklist, is to have the class negotiate about which of the main ideas and evidence should be part of a complete and scientifically defensible explanation of the phenomenon. Students will use this as a scaffold for writing their final evidence-based explanations in the last stage of the unit. There are a number of ways to facilitate this discussion and creation of the public record. We like to prompt groups to create a bulleted list of the three to five most important ideas that they think they need. We then ask for examples, press the whole class to make sure they understand how the idea fits in, and ask for consensus before writing it on the final checklist public record. This process can take about 15 to 20 minutes and most often goes quite smoothly by this point in the unit. However, this is also a time to make sure important ideas are included and to bring them up if necessary. Although this is uncommon, we may also provide some just-in-time instruction to tie up any loose ends in students' understandings from the unit.

The checklist becomes less useful to students if everything that comes up in this discussion is automatically written on the board. Ideas can generally be combined into five to seven bulleted points. We then go back and lead a discussion about the evidence we have for each of the bulleted points and write those alongside. This step is crucial, as students will need to coordinate the science ideas with evidence in the written evidence-based explanation in the next stage. At times, we have been unhappy with our checklist for some reason; perhaps the writing is not as legible or the points are not as clearly articulated as we would like. In such cases, we created a clean version that evening and asked the students the next day to ensure that it still represented all the ideas from the original poster.

Establishing Credibility Stage Summary

In the establishing credibility stage (Figure 4.9), students construct arguments in the form of evidence-based explanations of the phenomenon, which allows them to argue their ideas in writing. This stage also includes peer review and revision to establish credibility for their ideas. Before beginning this stage, review the relevant sections in Chapters 1 and 2 for specifics. We have provided a PowerPoint template to assist with this stage, which can be downloaded from the book's Extras page at *www.nsta.org/mbi-biology*.

Figure 4.9. Establishing Credibility Stage Summary

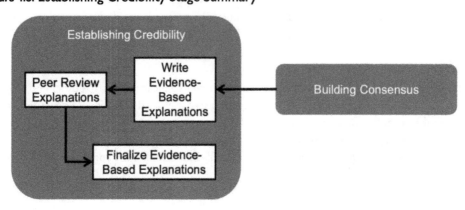

National Science Teaching Association

Write Evidence-Based Explanations

Up to this point, students have worked as part of a group to construct their evidence-based explanations of the phenomenon. Now, they work individually to write their final explanations. While they are on their own, they have a number of important public records such as the Summary Table and Gotta Have Checklist to use as references. Beyond these scaffolds, you may choose to provide additional support for students who may struggle with writing. For instance, we often have the first sentence written out on small slips of paper that we can hand to students to help build momentum. Some teachers choose to have student groups work together to create a group outline of their writing before students work individually. Your school may have writing support structures in place to help students get started. One thing we have found is that teachers are most often pleasantly surprised by their students' individual evidence-based explanations.

Conduct Peer Review of Explanations

Once students have finished writing their individual evidence-based explanations, it is important that they receive peer feedback before turning in their final product. Have students use the MBI Explanation Peer-Review Guide in Appendix B to provide feedback to each other. Beyond helping students improve their written explanations, this step reinforces the idea that in science, credibility comes not from an external standard but through negotiation with our peers.

Finalize Evidence-Based Explanations

After students have been able to review their peer feedback, and perhaps meet with their peers to discuss the feedback, they should use the critique to finalize their evidence-based explanations.

Evaluation

Once students have handed in their written explanations, you can use the MBI Explanation Peer-Review Guide in Appendix B to evaluate them. Besides clarity of communication, the guide focuses on whether a student's final product explains the phenomenon, fits with the evidence gathered throughout the unit, and builds on the important science ideas at the center of the unit.

Appendix A: Standards Alignment Matrixes

Standards Matrix A. Alignment of the Model-Based Inquiry Units With the Science and Engineering Practices, Crosscutting Concepts, and Disciplinary Core Ideas in *A Framework for K–12 Science Education* (NRC 2012)

	MBI Units			
SEPs, CCCs, and DCIs found in the *Framework*	**Unit 1. From Molecules to Organisms**	**Unit 2. Ecosystems**	**Unit 3. Heredity**	**Unit 4. Biological Evolution**
Science and Engineering Practices				
Asking Questions and Defining Problems	■	■	■	■
Developing and Using Models	■	■	■	■
Planning and Carrying Out Investigations	■	■	■	■
Analyzing and Interpreting Data	■	■	■	■
Using Mathematics and Computational Thinking	■	■	■	■
Constructing Explanations and Designing Solutions	■	■	■	■
Engaging in Argument From Evidence	■	■	■	■

Key: ■ = alignment.

Continued

Standards Matrix A. (*continued*)

SEPs, CCCs, and DCIs found in the *Framework*	MBI Units			
	Unit 1. From Molecules to Organisms	Unit 2. Ecosystems	Unit 3. Heredity	Unit 4. Biological Evolution
Obtaining, Evaluating, and Communicating Information	■	■	■	■
Crosscutting Concepts				
Patterns	■	■	■	■
Cause and Effect: Mechanism and Explanation	■	■	■	■
Scale, Proportion, and Quantity	■	■	■	■
Systems and System Models	■	■	■	■
Energy and Matter: Flows, Cycles, and Conservation	■	■	■	■
Structure and Function	■	■	■	■
Stability and Change	■	■	■	■
Disciplinary Core Ideas				
LS1: From Molecules to Organisms: Structures and Process	■			
LS2: Ecosystems: Interactions, Energy, and Dynamics		■		
LS3: Heredity: Inheritance and Variation of Traits			■	
LS4: Biological Evolution: Unity and Diversity				■

Key: ■ = alignment.

Standards Matrix B. Alignment of the MBI Units With the *Common Core State Standards for English Language Arts* for High School (*CCSS ELA;* NGAC and CCSSO 2010)

Grades 6–12 literacy in science and technical subjects	MBI Units			
	Unit 1. From Molecules to Organisms	Unit 2. Ecosystems	Unit 3. Heredity	Unit 4. Biological Evolution
Reading				
Key ideas and details	■	■	■	■
Craft and structure				
Integration of knowledge and ideas	■	■	■	■
Writing				
Text types and purposes	■	■	■	■
Production and distribution of writing	■	■	■	■
Research to build and present knowledge				
Range of writing	■	■	■	■
Speaking and listening				
Comprehension and collaboration	■	■	■	■
Presentation of knowledge and ideas	■	■	■	■

Key: ■ = alignment.

References

National Governors Association Center for Best Practices and Council of Chief State School Officers (NGAC and CCSSO). 2010. *Common core state standards.* Washington, DC: NGAC and CCSSO.

National Research Council (NRC). 2012. *A framework for K–12 science education: Practices, crosscutting concepts, and core ideas.* Washington, DC: National Academies Press.

Appendix B: MBI Explanation Peer-Review Guide

Explanation by: _____ Date: _____

Author

Reviewed by: _____ _____ _____

Reviewer 1 Reviewer 2 Reviewer 3

Section 1: Introduction	Reviewer Rating			Teacher Score		
1. Did the author provide an adequate **description of the phenomenon** of interest?	☐ No	☐ Almost	☐ Yes	0	1	2
2. Did the author make the **driving question** clear?	☐ No	☐ Almost	☐ Yes	0	1	2
3. Did the author provide an adequate overview of what they did to **figure out** the explanation?	☐ No	☐ Almost	☐ Yes	0	1	2

Reviewers: If your group gave the author any "No" or "Almost" ratings, please give the author some advice about how to improve this part of their explanation for the phenomenon.

Section 2: The Explanation	Reviewer Rating			Teacher Score		
1. Does the explanation include the **full causal story** of the phenomenon?	☐ No	☐ Almost	☐ Yes	0	1	2
2. Does the explanation include all the important **unobservable components** of the phenomenon?	☐ No	☐ Almost	☐ Yes	0	1	2
3. Does the explanation describe all the **relationships between the factors** involved in the phenomenon?	☐ No	☐ Almost	☐ Yes	0	1	2
4. Does the explanation include **essential science concepts** and relevant science ideas?	☐ No	☐ Almost	☐ Yes	0	1	2
5. Is the content of the explanation **consistent with the consensus model** developed in class?	☐ No	☐ Almost	☐ Yes	0	1	2

Reviewers: If your group gave the author any "No" or "Almost" ratings, please give the author some advice about how to improve this part of their explanation for the phenomenon.

Section 3: The Evidence	Reviewer Rating			Teacher Score		
1. Did the author support the explanation with **scientific evidence**?	☐ No	☐ Almost	☐ Yes	0	1	2
2. Did the author include enough **evidence to support the inclusion of each unobservable component**?	☐ No	☐ Almost	☐ Yes	0	1	2
3. Did the author include enough **evidence to support the proposed relationships between factors**?	☐ No	☐ Almost	☐ Yes	0	1	2
4. Did the author include enough **evidence to support the underlying mechanism**?	☐ No	☐ Almost	☐ Yes	0	1	2
5. Did the author do a good job of **explaining why the evidence** is important or why it matters?	☐ No	☐ Almost	☐ Yes	0	1	2

Reviewers: If your group gave the author any "No" or "Almost" ratings, please give the author some advice about what to do to improve the support for their explanation for the phenomenon.

Mechanics	Reviewer Rating			Teacher Score		
1. Grammar: Are the sentences complete? Is there proper subject-verb agreement in each sentence? Are there no run-on sentences?	☐ No	☐ Almost	☐ Yes	0	1	2
2. Conventions: Did the author use proper spelling, punctuation, and capitalization?	☐ No	☐ Almost	☐ Yes	0	1	2
3. Word Choice: Did the author use the right words in each sentence (for example, *there* vs. *their, to* vs. *too, then* vs. *than*)?	☐ No	☐ Almost	☐ Yes	0	1	2

General Reviewer Comments	Teacher Comments
We liked … We wonder …	

Total: _____/32

Appendix C: Safety Acknowledgment Form

I know it is very important to be as safe as I can during an investigation. My teacher has told me how to be safer in science. I agree to follow these 16 safety rules when I am working with my classmates to figure things out in science:

1. I will act in a responsible manner at all times. I will not run around the classroom, throw things, play jokes on my classmates, or be careless.

2. I will never eat, drink, or chew gum.

3. I will never touch, taste, or smell any materials, tools, or chemicals without permission.

4. I will wear my safety goggles, gloves, and apron at all times during the activity setup, hands-on work, and cleanup.

5. I will do my best to take care of the materials and tools that my teacher allows me to use.

6. I will always tell my teacher about any accidents as soon as they happen.

7. I will always dress in a way that will help keep me safe. I will wear closed-toed shoes and pants. My clothes will not be loose, baggy, or bulky. If my hair is long, I will use hair ties to keep it out of the way while I am working.

8. I will keep my work area clean and neat at all times. I will put my backpack, books, and other personal items where my teacher tells me to put them, and I will not get them out unless my teacher tells me that it is OK.

9. I will clean my work area and the materials or tools that I use.

10. I will follow my teacher's instructions for disposing waste materials.

11. I will use caution when using sharp tools or materials that can cut or puncture skin.

12. I will use caution when working with glassware, which can shatter and cut skin if dropped.

13. I will immediately wipe up any spilled water on the floor so it does not become a slip or fall hazard.

14. I will use caution when working with hot plates, which can cause skin burns or electric shock.

15. I will wash my hands with soap and water at the end of the activity.

16. I will follow my teacher's directions at all times.

_____   _____   _____
Print Name   Signature   Date

I have read and reviewed the 16 investigation safety rules with my child. My child understands how important it is to follow safety rules in science and has agreed to follow these safety rules at all times. I give my permission for my child to participate in the investigations this year.

_____   _____   _____
Parent or Guardian Name   Parent or Guardian Signature   Date

Image Credits

All images in this book are stock photos or courtesy of the authors unless otherwise noted below.

Task 11
Figures SH11.1 and SH11.2: Adapted from J. Roman and J. J. McCarthy, 2010, The whale pump: Marine mammals enhance primary productivity in a coastal basin, *PloS ONE* 5 (10): e13255, *https://journals.plos.org/plosone/article?id=10.1371/journal.pone.0013255*.

Task 17
Figure SH17.1: International Agency for Research on Cancer Working Group, 2006, The association of use of sunbeds with cutaneous malignant melanoma and other skin cancers: A systematic review, *International Journal of Cancer* 120 (5): 1120, *https://onlinelibrary.wiley.com/doi/epdf/10.1002/ijc.22453*.

Figure SH17.2: N. H. Matthews, W.-Q. Li, A. A. Qureshi, M. A. Weinstock, and E. Cho, 2017, Epidemiology of melanoma, in *Cutaneous melanoma: Etiology and therapy*., ed. W. H. Ward and J. M. Farma, chap. 1 (Brisbane: Codon Publications), *www.ncbi.nlm.nih.gov/books/NBK481862*.

Figure SH17.4: American Cancer Society, 2021, Key statistics for melanoma skin cancer, last updated January 12, 2021, *www.cancer.org/cancer/melanoma-skin-cancer/about/key-statistics.html*.

Figure SH17.5: C. G. Watts, M. Drummond, C. Goumas, H. Schmid, B. K. Armstrong, J. F. Aitken, M. A. Jenkins, et al., 2018, Sunscreen use and melanoma risk among young Australian adults, *JAMA Dermatology* 154 (9): 1001–9, *www.ncbi.nlm.nih.gov/pmc/articles/PMC6143037*.

Task 23
Figure SH23.1: NIH, Wikimedia Commons, *https://commons.wikimedia.org/wiki/File:06_chart_pu3.png*.

Figure SH23.2: Adapted from D. T. Zanatta, S. J. Fraley, and R. W. Murphy, 2007, Population structure and mantle display polymorphisms in the wavy-rayed lampmussel, *Lampsilis fasciola* (Bivalvia: Unionidae), *Canadian Journal of Zoology* 85 (11): 1171.

Figure SH23.3: Adapted from D. T. Zanatta, S. J. Fraley, and R. W. Murphy, 2007, Population structure and mantle display polymorphisms in the wavy-rayed lampmussel, *Lampsilis fasciola* (Bivalvia: Unionidae), *Canadian Journal of Zoology* 85 (11): 1173.

Task 28

Figure SH28.1: Virginia Department of Conservation and Recreation, Virginia's Major Watersheds (web page). See *www.dcr.virginia.gov/soil-and-water/wsheds* for the original, full-color version.

Index

Page numbers printed in **boldface type** refer to figures or tables.

Index

National Science Teaching Association

Index

National Science Teaching Association

Index

National Science Teaching Association

Index

National Science Teaching Association

Index

National Science Teaching Association